The Coming Storm

the Coming Storm

EXTREME WEATHER AND OUR TERRIFYING FUTURE

Bob Reiss

HYPERION
NEW YORK

Library of Congress Cataloging-in-Publication Data

Reiss, Bob.
    The coming storm : extreme weather and our terrifying future / Bob
Reiss.
        p. cm.
    ISBN 0-7868-6665-9
        1. Severe storms. 2. Global environmental change. I. Title.

QC941 .R43 2001
551.55—dc21   2001024377

Book design by Casey Hampton

FIRST EDITION

10 9 8 7 6 5 4 3 2 1

For a safe and beautiful world for Amit, Navi, Nicholas, Kareen, Noah Lev, Alec and Mitchel. And of course, Corky and Sandy. Clean air. Blue seas. White snow.

Acknowledgments

During the many months I spent traveling for this book and writing it, many people who do not appear in the pages generously shared their time, contacts and even their homes with me.

In Washington, D.C., many thanks to Jim Grady and Bonnie Goldstein, Barry and Karen Neuman, Ruth Ravenel, and Curt Suplee. In North Carolina, to Bill Choyke, Roz Ledford, Alan Basist, and to Phil Gerard, who took time from his busy schedule to read the manuscript and offer his usual sound advice. In Massachusetts, to Mel and Harriet Warshaw, and to Phil Shabecoff. In Nashville, Viola Ebert gave me the keys to her home and her car. In Oakland, Robert Bruce took time off from work to show me the city, share records and introduce me to victims of the 1991 fire. Richard Price of Chicago graciously opened his home to me, as did Bill Massey and Helen Marriage in London.

In New York, many thanks to Leigh Haber, a fine editor, and Peter Gethers, who helped conceptualize the project in early conversations. Esther Newberg, a fierce advocate and my beloved longtime agent, made the deal. New York assemblyman Mark Weprin was very helpful on the inner workings of government.

Dan Katz, past president of the New York–based Rainforest Alliance, helped arrange introductions to the Chiquita banana company, and in

Honduras, Jimmy Zonta of Chiquita helped arrange visits to the farms and interviews with executives and field workers.

UN Ambassador Tuiloma Neroni Slade of Samoa helped me contact Ambassador Hussain Shihab of the Maldives, who in turn approvingly relayed my request for interviews with President Gayoom back home to Malé. In the Maldives, Mohamed Khaleel made a reporter far from home feel perfectly at home.

Jerome Reiss, my dear old dad, lent me his new car in Florida. No matter how old you get, it's a big deal when your father lends you the new car.

And finally, in my own home, Wendy Roth put up with 18 months of often nonstop travel, and the loss of my usual (I hope) sense of humor during months of intense work. Wendy, thanks for helping tie down the kayaks when Hurricane Irene was coming. Thanks for the research. Whether the temperature is colder or hotter outside, you make our home warm.

The Coming Storm

Nashville

ARLY IN THE MORNING of April 16, 1998, a day anyone living in Nashville, Tennessee, at the time will never forget, a 52-year-old meteorologist named Bobby Boyd steered his Ford Takoma pickup through light dawn drizzle, along Shady Trail Road in Old Hickory, a suburb 15 miles east of downtown. He had a bad sinus cold but was heading for work. Unable to sleep the night before, he'd switched on his computer at home and downloaded severe weather forecasts from the National Weather Service's Storm Prediction Center in Norman, Oklahoma. He had not liked the reports.

"Outbreaks of severe thunderstorms with tornadoes, and damaging winds and large hail are expected for large parts of the Tennessee and Lower Mississippi Valleys," Bobby had read.

"The whole environment was unstable," he remembers. "There was severe weather to the west of us, in southern Illinois. Temperatures were going up, and when they do that at night in the spring, in advance of an approaching weather system, it's often a sign of impending bad weather."

Bobby Boyd knew that for Tennessee, in April, "bad weather" meant possible tornadoes.

Boyd is a slightly pudgy man—an ex–high school athlete who favors short-sleeve cotton shirts, khakis and soft brown chukka boots. Bald on top with short light-brown hair going gray at the sideburns, he speaks with a slight Tennessee twang, and is considered by his respectful colleagues to be thoughtful and professional.

By 1998 he had been a meteorologist for 30 years, and if tornadoes did touch down that day, his job would be to warn over a million people living in the middle and eastern parts of the state—including his wife and two daughters—in time for them to seek shelter.

Driving, listening to weather reports on his favorite country music station, he heard that the thunderstorms that had started to form in northwest Tennessee around 2:00 A.M. were still moving toward Nashville.

Boyd passed copses of cedar, hickory and birch as he turned onto a narrow road skirting Old Hickory Lake, named for Tennessee's most famous president, Andrew Jackson. He turned left at a sign reading "National Weather Service." An ascending two-lane blacktop brought him through a thickly wooded campground to a fenced-in compound marked by signs: "U.S. Property No Trespassing" and "National Oceanographic and Atmospheric Administration."

The rain was falling harder.

Boyd parked the Takoma opposite a one-story red brick building, the Nashville branch office of the National Weather Service. The small parking lot was dominated by a four-sided steel tower topped by a gigantic olive-colored glass sphere. Rotating inside, Doppler radar scanned surrounding skies for a radius of 125 miles.

Although it was early the parking lot was already filled, meaning extra staff had been called in—a logical step, Boyd knew, considering the rest of that weather advisory he'd seen at home.

"Conditions will be favorable for supercell storms and tornadoes, especially this afternoon through early tonight," it had said. "The threat for damaging winds and large hail will likely continue across much of the region through early Friday."

Boyd pushed through the glass doors and into the branch office, a bright, windowless, air-conditioned facility where meteorologists were issuing a severe weather warning for hundreds of farms, towns and cities in 43 counties stretching from the Tennessee River to the Cumberland Plateau—the whole middle part of the state.

He immediately began studying satellite reports, weather spotter reports and weather maps that had come in overnight, but was inter-

rupted by a call from a reporter who informed him that Tennessee's first
tornado of the day had touched down in Dickson County, well west of
Nashville, at seven. In the twister's short, violent ten-minute life, it had
cut a thirteen-hundred-yard swath along a two-mile-long area, destroying eight homes near Highway 46.

Three storm victims were on their way to the hospital. All would
live.

A long, unusual day was beginning, and would end up in the record
books.

Nashville was waking up. At the State Capitol, whose dome occupied
the highest ground in the city, legislators, aides and political journalists
were arriving at work. U.S. Interstates 65, 24 and 40 were clogged with
thousands of commuters. Nashville's Grand Ole Opry—home to the
country's best country-western recording artists—was quiet at this
hour. At the Tennessee Oilers' football stadium, the city Fire Department
headquarters, the Hard Rock Cafe and the Renaissance Hotel—none
of which would finish the day as intact as they were beginning it—
construction workers, visitors and businesspeople were discussing the
day's news.

Tennessee Democrats were considering running Lady Vols basketball
coach Pat Summitt for governor against incumbent Don Sundquist. A
state senate committee had voted 7–3 against reinstating the Racing
Commission.

What was *not* in the papers, considered less immediately relevant to
local interests, was an international fight raging over the possibility that
the earth's climate was changing, making weather more deadly across
the planet. That the "greenhouse effect," a heating of the earth accentuated by man-made gasses, was increasing incidents of hurricanes,
droughts, windstorms and tornadoes.

In fact, in Washington that day, the White House was backing a new
treaty to curb the greenhouse effect, calling for fossil fuel cutbacks, but
the treaty was opposed by Congress. Treasury Department experts
believed the treaty could cripple the nation's economy and fought
against it in the White House against economists who argued otherwise.
Powerful congressmen battled to keep the Clinton administration from

instituting energy cuts to put the nation in line with the proposed treaty.

Around the world, at the same time, a growing number of scientists, diplomats and business leaders were predicting that failure to curb the greenhouse effect could lead the world within 50 years into an environmental catastrophe, making the record-breaking weather disasters of the last few years small by comparison. Continue conducting business as usual, they warned, and we may push the planet across a critical threshold to a point where the atmosphere will violently change, causing rising seas, a surge in tropical diseases, mass migration from destabilized third world countries, killer heat waves and hundreds of small alterations in the daily lives of everyone on earth.

There was still time to prevent this cataclysm, they argued, but like the town alderman in the movie *Jaws*, who refused to believe a shark was offshore, their powerful opponents were putting the public in mortal danger by blinding themselves to the truth.

Meanwhile, political and economic interests opposing this view were spending millions to pay lobbyists, buy ads and fund scientists debunking the greenhouse theory. Whiny liberals were the ones putting the public in danger, they charged. Those "Chicken Littles," screaming that the sky was literally falling, wanted to change the way every country on earth does business—wanted to raise taxes and gas prices and monitor farmers—all because of some unproven scare theory that was nothing more than an excuse for conducting a vast social experiment and attacking Big Oil.

The believers had to agree that it was impossible to prove 100 percent that the greenhouse effect was changing weather. You could only prove it conclusively after it was too late, they admitted, although some in Nashville would come to believe that "too late," for them, came this April day.

So far, though, the rhythms of the day were normal.

At Vanderbilt University, the city's prestigious Harvard of the South, 22-year-old senior Kevin Longinotti was eagerly starting another busy day leading up to his graduation next month and the much anticipated beginning of his new life as a second lieutenant in the U.S. Army.

Longinotti had been up since before dawn, when he typically went through vigorous physical training with other ROTC students—students on army or navy scholarships. He was a quiet man but physically powerful, broad-shouldered, five-foot-nine and 165 pounds; a "lean, mean fightin' machine," his mom, Debi Slepicka, loved to joke.

Kevin was not the first person you'd notice in a crowded room, but teachers, ROTC officers and fellow students said he'd be the man you'd remember. His shyness hid a strong personality, active social conscience and impressive academic record. As far back as sixth grade, Kevin had asked his mom, "Where does the saying 'Knowledge is power' come from?" He'd been chasing knowledge ever since.

Now he was completing a triple major: education, mathematics and special education; taking 22 hours of courses instead of the usual load of 16. He was working as head resident adviser in a freshman dorm, Reinke Hall, where he was responsible for the safety of 90 male students.

"He got things done and he loved responsibility," said his best friend, Christine Pirozzi.

Kevin also tutored underprivileged kids at a local high school and had been named "superior cadet" in ROTC twice. But he was so modest even his friends did not know he'd won a prestigious Presidential Scholarship to the school, or that of the university's 5,900 students, he'd been the only one to win the award that year. They didn't know he'd also been accepted at Harvard and West Point.

"I suggested he go to Vanderbilt," says Debi, who preferred having her oldest son close to home.

It wasn't that Kevin didn't like to talk. Once you got to know him, everything was a story and he had to tell you right away. He spoke fast, using his hands to emphasize points. Every night he'd saunter out of Renke Hall and join up with four or five other RAs—his best friends—and they'd spread a blanket in the quad and talk for hours.

They'd discuss just about every damn subject they could think of, from the ozone hole in the atmosphere (Kevin stopped using aerosols as a result) to the sexual escapades of politicians (Kevin disapproved of them).

He was a joker too, good at keeping a straight face. One time, on a cold Christmas-break morning in Memphis, Debi was cold and she spotted a pair of Kevin's thick, clean black socks in the laundry room. Noticing them on her feet a few minutes later, he told her they were "special army socks, treated with chemicals to keep feet dry."

"Within ten minutes my feet broke out in hives," Debi told friends.

"Then the hives started spreading. Kevin started chuckling. There was no chemical on those socks."

Mom and son were close.

"He was afraid of only one thing in the world," Debi said.

Tornadoes.

"When he was twelve we lived in Little Rock, Arkansas. One day we heard the warning sirens go off. Kevin grabbed his little brothers and got them into the bathroom, into the tub. He dragged a mattress over them and lay on top to hold it down. We all crowded into the room. Then the power went out. We could hear debris hitting the house.

"He's been petrified of tornadoes ever since."

But on the morning of April 16, 1998, like most people in Nashville, Kevin was not thinking about tornadoes. Although the National Storm Prediction Center had announced they were "possible" today, the area throughout which they might appear was enormous, spread over several states, and the likelihood that any one particular person within that radius would actually encounter a twister was minuscule.

So, dressed in his usual tee shirt, jeans, baseball hat and flip-flops, Kevin went about planning a multitude of chores. He was scheduled to teach a disabilities class this morning at a local high school. He needed to get to First American Bank to make a payment on his '96 Altima. At some point, he had to study for his last final exam, scheduled tomorrow. Grabbing his car keys, he reminded himself he had one more event to attend this afternoon: the annual ROTC farewell picnic in Centennial Park, 12 blocks away—a fun-filled rite of passage, with volleyball games, Frisbee and flag football.

Elsewhere in Nashville, as Kevin left the dorm, Lieutenant Colonel Mike Patenaude, ROTC commander, was deciding if he should call off the picnic due to weather. At nine, at home, he watched the forecast on TV. A lean, balding man in his early forties, Patenaude was a veteran of service in Germany during the cold war, an ex-cavalry platoon leader who took his ROTC responsibilities seriously. He knew that even if he did hold the picnic he would miss its beginning because he had to take his daughter to the hospital. She was scheduled to undergo surgery on a knee injured during a soccer game.

Patenaude planned to come by the picnic—if it went ahead, that is— around four, after the surgery, and speak to the troops. First the 60 to 80 students would line up three lines deep, in platoons. Then Patenaude would call up Longinotti and the other seniors, announce which army branch they were off to join, and say goodbye. He'd call up the juniors, who were scheduled to attend a five-week-long army camp in Fort Lewis, Washington. He'd announce that they would be hungry at the camp, and tired. But he knew they would represent Vanderbilt well and make everyone proud.

The picnic would probably start breaking up about 4:30.

"In this part of the country, in the spring," he would later say, "thunderstorms and tornadoes are often possible. The morning had been rough. You don't want to go out in that stuff. I stayed home with my daughter and waited for the warnings to clear. Around ten-thirty I saw the line of storms on TV moving off to eastern Tennessee. I told my daughter, we can go to the hospital now. I even took her to Centennial Park and showed her where my people would have the picnic.

"It was beautiful outside, balmy, in the low seventies. People were jogging and walking and riding their bikes. A very nice day, mildly overcast, but no low clouds.

"I remember thinking, the picnic will be fine."

———————

After 30 years as a weather expert, Bobby Boyd knew that no storm on earth can produce winds as violent as a tornado—winds as fast as 300 miles an hour. Scientists cannot say *precisely* how high tornado winds rise, however, since the only way to know for sure is to place instruments directly in the path of the storm, an act which, even if you accomplish it, can get the instruments destroyed.

Tornadoes are so violent that they are rated partially by the kind of damage they do. An F5 tornado, highest category on the Fujita Tornado Intensity Scale, and the most feared wind on earth, can lift strong frame houses off their foundations and carry them hundreds of yards as easily as a left-fielder throws a baseball.

An F4, or "devastating tornado," packing winds up to 260 miles per hour, can lift a freight train, demolish well-constructed homes and toss huge uprooted oaks through the air as if they were packages of Hostess Twinkies.

A "severe tornado," or F3, will tear roofs off well-built houses with its 158- to 206-mile-an-hour winds. It will uproot trees or snap them off at the roots. It will toss Isuzu pickups into the air with the ease of a cook flipping flapjacks.

Tetsuya Theodore Fujita, inventor of this destruction scale, dates his fascination with damage to his early days as a researcher, when he surveyed the Japanese cities of Nagasaki and Hiroshima from the air, studying nuclear destruction. Appalled by the devastation, he dedicated his life to helping others prepare for disasters. Years later, at the University of Chicago, designing his labeling system, he decided that when a twister's winds reached 113 to 157 miles an hour, he would call it a "strong tornado," since it could still push over trains, blow cars off highways, tear roofs off frame houses and demolish mobile homes.

Even a "weak tornado," packing 73- to 112-mile-an-hour winds, peels surfaces off roofs, pushes mobile homes off foundations and can shove the family Ford into a ditch beside U.S. route 40 in Tennessee.

Bobby Boyd knew how to rate tornadoes, and he knew that of all places on earth, conditions favoring twister formation are most favorable in the U.S., in the spring, across a wide swath of plains starting east of the Rocky Mountains. "Tornado Alley" includes farmland, hills and valleys of Oklahoma, Texas, Kansas, Nebraska, Arkansas, Ohio and Tennessee.

Inside this area on April 16, 1998, three major conditions promoting tornado formation were visible on Boyd's weather maps. A cold dry mass of air, the first ingredient, was coming out of the northwest, having topped the Rockies by shedding its moisture. A midlevel jet stream would keep cool air close enough to earth so it would collide with warm air nearer to the ground.

The warm air, third ingredient in the deadly mix, was wafting in a wet, balmy tongue out of the Gulf of Mexico. It was a pleasant breeze and the moisture inside it was as potent as gasoline in a Molotov cocktail.

Bobby Boyd could see, on his weather maps, the area where the air masses collided. A vast funnel zone had been created where the fronts met, and the air inside had been set to slowly spinning. The hot air from the Gulf of Mexico was trying to rise, but the cold air above it contained it in the same way an iron lid keeps steam in a boiling pot.

Like the steam in the pot, the warm air was exerting increasing pressure on the lid above.

If the pressure grew too great, the warm air, pushing steadily upward in a spiraling helix, would smash through the cold air and break outward, at which point an awed pedestrian watching from earth, gripping his bicycle handlebars at a traffic light, or peering up from the passenger seat of a Honda, or even starting a fast run home over a wheat field to avoid rock-size hailstones beginning to fall . . . would see a huge, dark anvil shape above, a monstrously high cloud that, at its uppermost levels, contains ice.

Inside this huge cloud or "supercell thunderstorm," warmer and cooler air would form a violently rotating tube of air, invisible from the ground. High-altitude winds would spin the whole air mass, the rotation accelerated even further by more winds below.

The resulting storm could achieve an awesome beauty. The onlooker shielding his eyes and gazing up from the east or southeast, the only safe place to be during a tornado, would see, affixed to the bottom of the incoming supercell monstrosity, a floating mass of supercharged energy or dark "wall cloud" shaped like a black Tupperware bowl, as wide as a mile across, rotating like a UFO in a Spielberg movie.

Depending on the angle of the watching tobacco farmer, truck driver or softball player, the wall cloud could even take on a greenish tinge, as if pumping itself up with energy. Hail would begin plummeting from the north side of the storm. But the observer to the southeast, like a civilian watching the bombing of a city from a bluff, could bask in hot, bright sunlight at the same time.

Conditions would be perfect for a tornado to form. Bobby Boyd, at his radar screen now, envisioned what the watcher would see next. Suddenly a small nipple shape would form at the bottom of the bowl, and the nipple would elongate and descend in a tube shape.

From below, rising to meet it, would come a swirl of dust and debris as the tornado's strength and intensity became visible and the funnel cloud linked sky to earth.

The thunderstorm would have become a floating power pack for its tornado.

Boyd knew that rotating debris in a powerful tornado can go up 1,000 feet. He knew that during a tornado's brief average life span of between 8 and 10 minutes (big tornadoes have stayed on the ground as long as 45 minutes) it can cut a swath of destruction a mile across.

He knew that despite its destructive power, a tornado—the rotating wall cloud, the pulsating colors, the gossamer cloud veil extending miles to the northeast of the storm—could transfix experienced storm chasers, freezing them in place like a rodent in front of a cobra.

"The beauty can be hypnotizing."

————

But Bobby Boyd had been trained to not even hesitate if a tornado showed up. Thousands of lives would depend on the timeliness of his warnings today. If he issued one, schools would empty. Airports would close. Farmers would herd their families into storm cellars. Hundreds of thousands of curious citizens would be drawn against their better judgment to their windows, where they would peer out and wait for a close-up view.

Boyd's own view would be quite different from one at a window. If a tornado appeared, he would see a computerized vision, a radar picture or satellite shot from 23,400 miles over the equator, rolling out on an overhead monitor.

Boyd's workplace was bright with artificial light and perpetually air-conditioned for the benefit of the sensitive machinery inside. It was glassed-in, so an observer in the hallway outside could watch the intense activity on the other side of the pane. Meteorologists received steady updates from the Storm Prediction Center in Oklahoma. They typed standardized weather warnings and dispatched them with the punch of a button. They answered a stream of phone calls from reporters and Tennessee state Emergency Management Agency staffers who wanted updates.

At 9:18 that morning, those updates had included the fact that a second Tennessee tornado of the day had touched down in Montgomery County, north of Nashville. During its violent 22-minute life, it had traveled eight miles, cut a swath 400 yards wide, knocked over a TVA transmission tower and ripped part of a roof off Jo Byrns School in the town of Adams. It left $410,000 worth of property damage.

At least no one had died.

At 10:45, when Mike Patenaude was driving with his daughter toward Centennial Park, the monitors showed a clear area over Nashville, but a line of gray swirling masses—the next round of storms—approaching from the west.

At 11:30, a two-note alarm went off on Boyd's Doppler radar unit. A "mesocyclone," or powerful thunderstorm severe enough to form a tornado, had been detected northeast of the city, in Macon County. Several minutes later it produced an F2 tornado that started snapping off trees and power lines along a swath 800 yards wide.

No one was hurt.

Now it was noon and though Boyd was feeling ill he remained at the Doppler radar unit, one of the room's most crucial pieces of equipment. It is so important that two meteorologists always man it when twister touchdowns are possible, in case one looks away, or goes to the bathroom, or fails to hear the alarm go off.

Doppler radar can actually spot a tornado forming inside a thunderstorm *before* it becomes a tornado. It can "see" the rotating wind. Unlike traditional radar, which measures the reflectivity of what it hits—rain in a thunderstorm for instance—Doppler radar can also detect whether the rain is blowing toward or away from the unit.

Since rain inside a mesocyclone blows toward *and* away from a viewer at the same time as the storm rotates, the Doppler "sees" the speed of the wind and whether it is approaching twister intensity.

Like the real tornado, the image is lovely to watch. On Doppler, a storm appears as a gorgeous bouquet of colors, as vivid as any in a prized canvas in a museum of art.

And as in a Seurat Impressionist masterpiece, the overall picture seems solid from a few feet away but is actually comprised of hundreds of tiny pixels, or colored dots. In the painting, when enough dots cover the canvas, they merge into a picture and a park appears. An art lover sees not dabs of paint but men in top hats and women holding parasols. There's a green meadow and a blue pond.

Only when the viewer gets close to the canvas does he realize he's seeing a myriad of individual dabs of paint.

On Doppler radar the same principle applies to rain inside a thunderstorm. Rain blowing *toward* the unit appears as tiny rectangular pixels ranging in color from the light green of the Augusta Golf Course fairway to the dark hues of an Oregon forest after a storm. If the rain is falling lightly, the Doppler shows individual pixels. But if the rain is driving, the pixels merge into solid green.

Rain shooting *away* from the unit shows up as pixels the bright happy red of Santa's coat.

An experienced Doppler watcher knows that if masses of red and green appear beside each other, the wind inside the storm is rotating. Rain is blowing *toward* the radar on one side of the funnel and away on the other.

If the wind intensifies to tornado intensity, at the core of the funnel, the red bleeds into a watery pink, and the green shifts to a sucked-out blue.

On screen, a little red triangle appears.

And Bobby Boyd cries, "We have one!"

———

By 12:30 Nashville was sunny and warm. At Vanderbilt University, the ROTC students were getting ready for their picnic. Mike Patenaude was at the hospital with his daughter. Kevin Longinotti was making a payment on his Altima. Bobby Boyd, 15 miles to the northeast, punched a switch on the Doppler unit to give him a different view of a tornado should one occur.

Now he had the unit in "reflectivity" mode, to give him a wider view of the surrounding area. His gray-black screen showed the boundaries of Tennessee counties outlined in pale white. Masses of rain were superimposed over the screen in colors corresponding to their density. If a supercell storm floated into radar view, Boyd would see a vaguely triangular mass stretching out in an anvil shape on its northeast side. The fringes of the storm would be corncob yellow, where rain fell more lightly.

The center would be fire-engine red.

Most alarmingly, in the southeast, the whole mass could condense into a curving tube extending from the wall cloud, appearing as a small, bright red hook.

If that happened, Boyd would cry, "We have one!"

———

Centennial Park is a well-loved oasis of recreation in west Nashville. It is bounded by Charlotte Avenue, a wide, main commercial thoroughfare leading to the capitol, and by West End Avenue on the south. The park features a scale replica of the Parthenon, which houses an art museum. There's a small pond, and benches for sitting, and a meadow for playing softball or Frisbee. An old F-86 Tennessee Air National Guard fighter jet is mounted on display.

On the north end of the park, at 2:30, the ROTC group was beginning their picnic by a large open-air pavilion and a row of big 150-year-old oaks. A thickly wooded bluff blocked the view to the west, the direction from which weather was approaching.

"After a while the sun went away but it was still warm," recalls Dave Burnstein, who was in the park as an ROTC sophomore that day. "People were eating hot dogs and burgers. The grill was going. The coolers were full of soda and pretzels. Some guys were throwing a football."

At the same time, Bobby Boyd was watching a line of potent thunderstorms approaching Nashville.

"By two-twenty they'd reached Dickson County. At two-thirty-five we put out a tornado warning for Dickson. I had sufficient depth of the mesocyclone on the Doppler to get reports of touchdowns, but I didn't have any yet."

By 3:04 the storms had reached Cheatham County, next county over from Nashville. Boyd spotted the hook-shaped appendage he had been fearing. His partner Mike Murphy instantly started typing out a tornado warning for the area as Bobby kept his eyes on the screen.

"Be sure to include downtown Nashville in the warning," Bobby said.

At 3:09, Murphy hit a button, and the warning went out onto the National Weather Service's public telephone line.

"A tornado will be passing through Nashville around three-thirty," it said.

Boyd took a quick break and called home to warn his wife. But she was out. "A tornado is approaching Nashville," he told their answering machine.

He also called one of his two daughters at her job at Wheels & Deals, and told her, "Get the radio on."

He tried to reach his other daughter, and failed.

"I kept thinking about 1933, when a similar supercell had hit Nashville and killed twelve," he said. "I kept seeing the image of that destruction in my head.

"At the office, we had done all we could do."

If Nashville had a tornado early-warning system, like many towns in Tornado Alley, at this point sirens would have started wailing all over the city.

But Nashville lacked a tornado warning system, so the old Civil Defense Air Raid sirens around the city remained silent. Bobby Boyd's warnings were being broadcast urgently over local TV and radio, but if you had no access to TV or radio during that half hour, you were out of luck.

Mike Patenaude was still at the hospital, by his daughter's bedside. The surgery had not begun yet, and Patenaude was looking up at the TV.

A few minutes earlier he'd heard the announcer say a "storm" was approaching Nashville, but no tornado alert had been issued. From the red arrows on screen, it seemed to Patenaude like the storm would veer north of the park.

"I thought, I wonder if it will affect my people."

Meanwhile, *inside* the park, Dave Burnstein recalls, "the wind started picking up, blowing paper plates around. It looked like it was going to rain. It was turning bad fast. We started picking up the trash. I looked up, toward the bluff, and saw a black cloud coming down off the hill. It was a different color from the rest of the sky. I thought, what the hell is *that*?

"I don't remember who started yelling, '*It's a tornado!*' Five seconds later it was here. I dropped my trash bag and ran to one of the poles holding up the pavilion. I wrapped my arms around the pole. Some people were trying to run. I saw things flying. The fighter plane on the pedestal was vibrating so hard I expected it to take off. I held on to that pole. The big trees all around the pavilion, hundred-and-fifty-year-old trees, were being ripped out of the ground, or snapped off at ground level.

"People were screaming, '*Oh my God!*'"

In Memphis, at the Bel-Air Bar & Grill, Debi Slepicka had been serving customers at the bar; chatting and laughing with them. Shortly after 3:30 she suddenly froze.

"I had a premonition about Kevin."

Debi rushed to the phone and called Vanderbilt. Sure enough, the operator who answered told her, a tornado had hit Nashville, but the students were safe, in basements.

Debi went back to work.

"I felt a little better, but I knew I needed to talk to Kevin before I could relax."

In Centennial Park the tornado was over. The students who had wrapped themselves around the steel poles of the pavilion let go. The ones who had taken cover in a ditch crawled out. They all stood in the rain, aghast, staring at the immense trees that had been ripped from the ground.

At first it seemed like everyone had escaped injury.

Then someone noticed a body beneath one of the fallen trees.

Kevin Longinotti, still conscious, lay in the mud, his hip and pelvis pinned by the oak.

He looked up. His voice was steady. He did not seem panicked.

"Would someone get this damn tree off me," he said.

The tornado, spinning at F2–F3 force, was now moving east along wide, commercial Charlotte Avenue, toward the State Capitol. At the American Red Cross, a wall blew in. The tornado blew the clock off the tower of the Union Station Hotel. It destroyed a wall in the Sheriff's Office. It ripped part of the roof off the capitol, and buried valuable firefighting equipment under debris at the Metro-Nashville Fire Department headquarters. The Tennessee Oilers' football stadium, one-third complete at the time, suffered damage.

The storm peeled a mural off the Hard Rock Cafe downtown and tore apart an elevated glass-enclosed crosswalk connecting the Renaissance Hotel and the Church Street Shopping Center.

Hundreds of homes were destroyed or damaged in east Nashville. Over a thousand trees were blown down at the Hermitage, historic estate of former U.S. president Andrew Jackson.

The twister, almost a mile wide, was now heading out of town and into the suburbs, toward Bobby Boyd's office.

As it approached the National Weather Service Operations Center, Boyd's boss, Derrel Martin, was on the phone being interviewed live by an NBC radio reporter in New York. Derrel is a big, broad-shouldered, white-haired man who dresses in western clothes and boots and likes turquoise jewelry. His good-natured humor contrasts directly with his imposing appearance. Like Bobby Boyd, he had over 30 years experience on the job.

The reporter was saying, "Will you excuse me while we go on a brief commercial break?"

Derrel, staring out the window at flying debris on the horizon, answered cordially. "Yes ma'am, you can be excused. But I won't be here when you come back."

"Why is that?" the reporter asked.

"Because," Derrel said, now hearing the approaching roar, "there's a tornado half a mile away."

When the tornado hit Centennial Park, only a few blocks away from one of the South's finest hospitals, "We were geared up for Armageddon," a Vanderbilt University Medical Center spokesman later remembered. "We'd gotten a call from the Tennessee Emergency Management Agency telling us to expect hundreds of casualties, so we kept our first-shift people here, and the second shift was coming in. We brought patients away from windows, into the middle of their rooms. We set up a command center in the admitting office.

"A little while later the sky turned green-black. The air got warmer and the leaves stopped moving outside. We were in radio contact with the far side of the campus, where people could see a tornado going by.

"After that the sirens started."

Inside the Center, Dr. Tim Van Natta, a surgeon and trauma expert, waited for the flood of injured with the rest of the staff. Van Natta, 38, was a lean and fit long-distance runner, originally from Seattle. Son of a U.S. Forest Service worker, he'd brought his wife and two children to Nashville a couple of years earlier, drawn to the quality of the medicine at Vanderbilt and the challenge of trauma work.

On April 16, 1998, Van Natta, unable to contain his curiosity, had been drawn to the window of the trauma unit when he heard that a twister might pass. He'd even seen the funnel fly directly over the roof of the Emergency Room Hospital a couple of blocks away.

"It wasn't on the ground yet."

Soon after he learned that the first victim of the tornado, a 22-year-old ROTC student named Kevin Longinotti, had arrived at the hospital. Longinotti's only piece of good luck so far today had been that a construction crew working in Centennial Park had been able to use its heavy machinery to lift the oak off the boy.

"The tree had crushed his pelvis and upper leg," Van Natta said. "And he was bleeding massively from the pelvis."

Van Natta knew he had to work fast. Usually the hospital gets about five patients a year who die from pelvic fractures, generally after a car accident.

"We're unable to stop the bleeding in those cases."

In Kevin's case, "the evaluation in the ER took about half an hour. The pelvic fracture was the most severe. Kevin also had a break on the upper part of his leg, and we found interabdominal injuries on the CAT scan. If we hadn't diagnosed those he could have gotten peritonitis. We took him to the operating room and opened his abdomen. He was bleeding into it. We started 'packing,' putting sponges in to soak up the blood. Also, his left colon above the rectum had been torn, so we had to remove part of that."

But the big problem was the pelvis. Kevin's injury was different from most. The majority of pelvic injury patients Van Natta sees, from car accidents, "are suffering from a springing-open of the pelvis," Van Natta said.

"We can put 'em back together and stop the bleeding in most cases. Usually we replace the lost blood and the clotting factors . . . Sometimes orthopedic surgeons will build, like an erector set, something called an 'external fixator' to compress the pelvis.

"But Kevin's pelvis had been smashed by the tree, so it wasn't amenable to that."

He was still alive, though, heavily sedated, and after several hours with Van Natta in the operating room, he was in stable condition, for the moment at least.

———

At 6:30, in Memphis, the phone rang at the Bel-Air Bar & Grill. The caller asked the bartender if he could speak to Debi Slepicka.

"Are you Kevin Longinotti's mom?" the man asked when she got on the line.

She started shaking. She handed the phone to someone else.

"I couldn't listen to what he had to say."

———————

By 11:00 P.M. when an exhausted Bobby Boyd was home, the most costly and destructive tornado outbreak ever to hit Tennessee was over. Boyd and his fellow meteorologists had issued 107 tornado warnings and 54 severe thunderstorm warnings since dawn, for the 44 counties in the area covered by the Nashville office.

Sixteen tornadoes had touched down in the area, and two of those had passed close to Boyd's office. But thanks to the timely warnings only 7 people had died across the state, and only 50 had been hospitalized in Nashville. The vast number of these injuries were not life threatening, although two patients' lives hung in the balance.

"Providence was smiling that there were not more injuries," Mayor Phil Bredesen told his stunned city.

Downtown Nashville was closed. In the hardest-hit East End, hundreds of homes were destroyed. Thousands of people were homeless. Damage was estimated at over $100 million.

Governor Don Sundquist declared the city a disaster area, and federal and state Emergency Management Agencies were preparing to rush personnel to the area, as were the Red Cross and Salvation Army.

Bobby Boyd, finally lying in bed after almost two days without sleep, knew that April traditionally has been the worst month in Tornado Alley for tornado deaths. That a day like today had happened in April was considered within the natural range of tornado activity by many meteorologists.

Even though by April 1998 over 100 people had been killed by twisters that year—three times the normal number—across the country, many experts regarded this as part of "normal" weather fluctuation. Some years there were more tornadoes and some years there were less. Plus, in today's case, the right atmospheric conditions for a tornado outbreak had existed. These conditions had caused tornadoes to form for millions of years.

But Boyd also knew that in 1998, some meteorologists were starting to believe that there was something "different" about this year's storms.

That rising global temperatures—caused by the greenhouse effect—were changing weather patterns in a way that might have influenced today's tornado outbreak.

Gary Grice, deputy director of the National Centers for Environmental Prediction's Storm Prediction Center in Norman, Oklahoma, said, of 1998 tornadoes in general, "It's scary. They seem to be stronger and hitting in the wrong places."

Boyd himself doubted that the greenhouse effect could have influenced the tornado. Derrel Martin, his boss, had a different opinion.

Either way, on this evening, while Kevin Longinotti's life hung in the balance, the debate over the greenhouse effect had become much more than a scientific exercise. It had entered the realm of global economics and politics.

By April 1998, each time any extreme weather event occurred, it became fuel in the steamy debate over climate. Was a particular storm simply part of natural variation? Or had it been made worse by the greenhouse effect?

Nineteen ninety-eight would turn out to be the hottest year on record—at least until 1999—and would provide many record-breaking events to argue about. In China, 240 million people would flee an area six times the size of Holland when the Yangtze River flooded after severe rains, in a "flood of the century." Hurricane Mitch, a "500-year storm," would kill 10,000 people in Central America.

In the U.S., monthly rainfall records would be broken in New Orleans; Asheville, North Carolina; Burlington, Vermont; and also in 19 places in California. Fifty-four of Florida's 67 counties would be declared federal disaster areas.

With each disaster the arguments grew more severe. At international conferences Saudi Arabian delegates shouted down British delegates. Kuwaiti representatives argued bitterly with the Dutch. China and India accused wealthy countries of trying to use greenhouse theory to keep them from developing. Frightened island countries begged for aid against rising seas.

In Nashville, though, Debi Slepicka stayed by Kevin's bedside each day and spent nights upstairs in a conference room into which nurses had wheeled a bed.

*"I couldn't sleep," she remembers.

The long table in the room was piled with food and drink brought by friends, relatives, well-wishers.

"I couldn't eat."

As Kevin Longinotti, unconscious and heavily sedated, lingered for days, Vice President Gore called the hospital to check on his condition.

Country-western star Trisha Yearwood made an appearance after receiving a letter from one of Kevin's friends, telling her he was a rabid fan. Trisha sat on his bed and ruffled his hair. He remained unconscious.

There was a small notebook beside his bed, where visitors recorded messages. One said, "Kevin . . . you're one little tough Longinotti, so hang in there and keep fighting for your dad. We love you and pray for you. I love you."

Another notation said, "Hey, Big Man . . . You just keep beating the odds. Keep pullin' and drivin' on . . ."

Each time Dr. Van Natta checked his patient there were more notations in the book. If the wishes were powerful medicine Kevin would have been cured. But even after the early operation there was a problem with pelvic bleeding. Lab tests showed an elevated myoglobin count, which meant Kevin's muscles were breaking down. He was strong, fighting to live, but his periods of stability were getting shorter. His blood pressure was sinking, his heart rate rising. He was developing septic shock.

Van Natta conducted a series of operations to try to remove Kevin's dead muscle without amputating the leg. For eight or nine days, doctors brought Kevin in and out of the operating room, cleaning out the wound, removing dead muscle. "Once muscle is dead, bacteria jump onto it, see it as food," Van Natta said. "After that, it's a race between us and the bacteria. Eventually it came down to a point where they talked about doing a hind-quarter amputation. Removing the entire left leg, hip and half the pelvis. It would have been a massive mutilating amputation, and his chances of survival would have been less than ten percent after that.

"But his infection was just so overwhelming. We couldn't maintain his blood pressure. He was on very high ventilator settings. He had a lot of trouble because with the shock the kidneys fail, the liver fails, the lungs fail. You have multi-organ failure. The heart doesn't work as well. The blood system fails. The lining of the abdomen fails. The bowel fails.

"Everything fails."

———

Kevin Longinotti graduated posthumously, with honors, a victim of weather during a day rated by the Tennessee Emergency Management Agency as having hosted the most costly, destructive tornado outbreak ever to hit the state. If greenhouse theorists are right, his death is a harbinger of the kinds of tragedies that will occur with greater frequency in the future. His and thousands of other deaths due to severe weather over the last decade provide fuel to the growing debate.

Longinotti had probably never heard of the man who had launched this debate onto the global stage ten years before, a quiet scientist named Jim Hansen. The name was only vaguely familiar to Bobby Boyd. Debi Slepicka said she had never heard of Hansen a year and a half after the tornado, on a November day, when she sat in a backyard in Memphis, showing off photos of Kevin to a reporter.

She still had a hard time sleeping, she said. Sometimes she drove to Nashville and walked around the Vanderbilt campus, because she felt closer to Kevin there. She'd visit the ROTC building and read the plaque commemorating his death. She was campaigning in the press to get Nashville to install a tornado warning system.

Eventually that late afternoon, Debi asked the reporter why he was interested in the tornado anyway.

He explained that he was writing a book about severe weather around the world . . . record heat waves and storms that seemed to occupy more space in the headlines. He was trying to get past the rhetoric and learn if the climate was really changing—meeting victims of weather and tracking the science and politics of an issue that could very well be the most crucial of the twenty-first century. The book would cover the period between 1988, when Jim Hansen shook the world with his announcement, to 2000.

"Well, I know what caused the tornado we had in Nashville, and it was the greenhouse effect," she said, surprising the reporter, who had not mentioned it.

There was no doubt in her mind. She was not a scientist, just an average person. But she reflected a growing consensus and political force.

Then Debi asked who Jim Hansen was anyway.

"How *did* he shake the world up in 1988?" she said.

Washington ... Wyoming ... Antarctica

SHORTLY AFTER LUNCHTIME on June 23, 1988, one of the hottest, most humid, most miserable days in Washington, D.C., that year, clerks readying for a Senate hearing—testing microphones or filling water pitchers in room 366 of the Dirksen Senate Office Building—realized something big was afoot. Reporters from the *Washington Post*, *New York Times* and the national networks were gathering in the hallway. Lots of hearings are held in Washington but most don't attract so much attention.

The reporters had been alerted by Senator Tim Wirth's office that a scientist named Jim Hansen would have something startling to say today.

Hansen, at the moment, was making his way to Capitol Hill in the back of a taxi. He headed NASA's Goddard Research Institute for Space Studies, in New York City, where he studied earth's climate.

Lately, he had been troubled by what he'd found.

Visibly, there was nothing imposing about the man who would start an international controversy that day. Jim Hansen radiated a mild, shy, collegiate aspect, like Jimmy Stewart in *Mr. Smith Goes to Washington*. He wore a tie and jacket but was more likely to dress in argyle vest sweaters, matching socks and high tops or sandals in the office. His high forehead and thinning straw-colored hair gave him a vulnerable look, but it was deceptive.

Hansen says what he thinks.

Still, testifying before a Senate committee was probably the last thing on Hansen's mind when he'd grown up outside tiny Denison, Iowa, where signs greet travelers with: "IT'S A WONDERFUL LIFE. HOME OF DONNA REED."

Son of a tenant farmer and a waitress, the future target of White House Chief of Staff John Sununu's fury was a lackadaisical high school student, more interested in playing second base in Babe Ruth ball than reading science. But he'd used money saved on his paper route to help pay for college at the University of Iowa, where he'd become fascinated with physics and astronomy, and ended up earning a BA with the highest distinction in physics and math.

After working as a graduate assistant at the Institute for Astrophysics in Kyoto, Hansen had written his PhD dissertation on the atmosphere of Venus, where concentrations of carbon dioxide are so high that the greenhouse effect—the phenomenon by which atmospheric gasses trap the sun's heat and help warm their planets—raises surface temperatures to 800 degrees Fahrenheit, hot enough to incinerate a sirloin.

"The study of other planets is very appropriate for understanding things that affect temperatures on earth," said Hansen, who later became principal investigator of experiments on the Pioneer Venus and Galileo Orbiter satellite projects.

In fact, if Venus was broiling due to so *much* carbon dioxide in its atmosphere, Mars had so *little* compared to earth that almost all of the sun's energy reaching the red planet was radiated back to space. Mars's surface temperature, thanks to its small greenhouse effect, was only a few degrees above what it would be if its atmosphere contained no carbon dioxide at all.

"On earth, the natural greenhouse effect keeps temperatures about thirty-three degrees warmer than they would be otherwise," Hansen liked to say.

Meaning, earth's natural greenhouse effect had created conditions enabling life here to thrive.

But Hansen had come to Washington today to talk about whether changes to the natural greenhouse effect, caused by humans, were dangerously altering the earth's climate. It was a particularly relevant question to the waiting reporters because, as any of them could see outside, or hear in their newsrooms, the whole country seemed to be burning up. In the capital, the temperature was 98, three degrees down from a record-breaking 101 the day before. After two weeks of nonstop heat wave, the local power company, Pepco, was strategically cutting service to keep Washington provided with electricity. To the east, Maryland's corn and soybean crops were baking. To the west, the wheat crop was dying across the Dakotas and Minnesota. In Montana, the earth was packed so dry from drought that ranchers reported using dynamite in places instead of shovels to dig fence posts. And to the south, in Memphis, the Mississippi River had sunk to an all-time low. The Mississippi was so shallow that 23 tugs towing 460 barges had run aground in the last week, along the stretch linking Rosedale, Mississippi, with Cairo, Illinois.

From Georgia to Arkansas to Texas, meteorologists were calling the drought the worst since the Great Depression, and now, with minutes to go until the hearing began, in the Dirksen Building, freshman senator Tim Wirth of Colorado was thinking that politically at least, the drought was a great thing. The first-term senator, a tall lanky liberal and ex-congressman, had actually *tried* to schedule this hearing during a heat wave, to attract more attention to what Hansen and other scientists would say.

"I'd had my staff call the meteorological people and ask, statistically, what's the hottest time of year in Washington. I also had the clerks open the windows that day," Wirth would remember gleefully, years later.

The journalists and senators trickling into the room—Max Baucus from drought-ridden Montana, Dale Bumpers from parched Arkansas, J. Bennett Johnston of Louisiana, where much of the coast area lay vulnerable to storm surges—had already heard of the greenhouse effect. A few had even heard of a Swedish chemist named Svante Arrhenius who'd come up with an unnerving theory about it as far back as 1896.

Arrhenius, sitting back in Stockholm early in the industrial revolution, gazing at smokestacks dotting the pollution-darkened sky, had started multiplying in his mind all the coal he imagined being burned in Scandinavia and London and Paris and New York, all the millions of tons of carbon dioxide from wood fires and coal furnaces, being pumped into the atmosphere, creating energy to drive ships or heat bathwater or to power the vast, growing industrial complexes churning out steel and textiles to make life better for millions throughout the world.

Among the consequences of this wonderful progress, Arrhenius envisioned a day in the future, long after he was dead, when so *much* man-made carbon dioxide would flood the skies that, in the same way too much of any foreign chemical changes the composition of a finite area—a pond or river, where a toxic substance is dumped—the extra carbon dioxide *would begin altering the atmosphere of the earth.*

Of course, the atmosphere was much vaster than any pond, and the process would take much longer, and of course Arrhenius's disturbing notion was only a theory, but if it *did* come true, the Swede predicted, climate would begin to warm. The extra carbon dioxide would create a heightened greenhouse effect, blocking more of the sun's energy from radiating back to space after it reached earth—trapping extra heat on the planet. And since heat basically drives all weather on earth, *extra* heat might push the whole climate system out of whack, the theory went, worsening the severity of wind, storms and droughts.

Of course science, in 1896, was filled with lots of theories that turned out to be false. But Arrhenius's didn't die. As Jim Hansen and Tim Wirth knew, by the last quarter of the twentieth century, readings taken by scientists at the Mauna Loa Observatory in Hawaii had proved that carbon dioxide content had risen in the atmosphere, just as Arrhenius had predicted, from 314 parts per million in 1958, when measurements began, to almost 360 ppm, a 15 percent growth.

Would the climate also warm as Arrhenius had predicted?

Worried about consequences, the National Research Council had formed a study group on climate in 1979 and stocked it with experts from Harvard, the University of California, MIT, the University of Stockholm, the National Center for Atmospheric Research and the Woods Hole Oceanographic Institution. The group was asked to assess the possibility that the extra CO_2 could cause the climate to change.

Their report had said, "It appears that the warming will eventually occur." And that "associated regional climatic changes might well be significant."

By today's hearing, Wirth had learned from researchers at the National Center for Atmospheric Research exactly what the "associated regional changes" might be. With doubled CO_2 in the atmosphere from pre-industrial times, a possibility predicted by the year 2030, oceans could rise due to melting polar ice and thermal expansion (the process

that makes water rise in a boiling pot). The size and intensity of hurricanes would increase. Since heat evaporates water, and evaporating water
returns to earth as rain, areas prone to flooding would suffer more of it
under a heightened greenhouse effect. Drier areas like the Great Plains
or the sub-Saharan region of Africa would become more susceptible to
drought.

The predictions encompassed every area of life. The changes would
start slowly and accelerate. Whole ecosystems might perish as a new
climate destroyed them. Millions of third-worlders would flee from
storm- or drought-ravaged Pakistan, Sudan, Central America, to Europe
and the U.S. Billions of dollars would be needed for sea dikes. Tropical
diseases like malaria would move north.

"We only have one planet. If we screw it up, we have no place else to
go," Senator Bennett Johnston, who'd read the reports, planned to tell
the committee today, when the hearing began.

On the other hand, the worst-case scenarios hadn't happened yet,
and even most scientists who worried about the greenhouse effect in
1988 merely said it "might" change the climate, "might" cause more
storms, "might" be responsible for the observed heating so far.

Jim Hansen walked into the room.

On this day, his words would drive the old theory out of the realm of
pure science and into power politics. Today's hearing would turn out to
be much more than an intellectual discourse. To Wirth and his powerful
ally, Senator Al Gore of Tennessee, it was time to take action *before* things
went out of whack and while any changes were still small and reversible.
If human-made carbon dioxide emissions were damaging the earth's
climate, Wirth and Gore figured, they must be cut. It was time for a
frontal assault on coal and oil usage—by law, by global treaty and even
by extra taxes or subsidy cutoffs to Exxon and Mobil and the other big
energy and auto giants.

Not so easy to do.

As the hearing was called to order, opening guns were sounding in a
public, 13-year battle over the way the U.S. and the world uses energy.
Just tracking the issue's journey so far was a study in how topics move
from chitchat in senators' offices to front pages of the *New York Times*.

In choosing climate to start with, Wirth had been looking for a
"horse," an issue on which he could become a prime mover, said David

Harwood, a key aide. Wirth, an environmentalist, had been provided with the issue when a World Resources Institute senior associate named Rafe Pomerance, a big, excitable, well-liked Washington environmentalist, had arrived in his office one day, waving a climate report.

"Rafe walked in with that wonderful smile, jacket hanging off one shoulder . . . I didn't know him from a hole in the wall, and he's saying, '*I gotta talk to you about something!*' " Wirth said.

Wirth didn't know much about the greenhouse effect at the time, but "fortunately, in the district I represented we had the National Center for Atmospheric Research. And Steve Schneider, one of the leading climate scientists in the world, was a neighbor. The scientists would come to dinners at our house. We'd see 'em all summer long."

Wirth began following the issue.

When he reached the Senate, taking over disgraced Coloradan Gary Hart's seat, the next step had been coordinating climate strategy with fellow Democrat Al Gore. The two men had discussed it during a car ride.

We were "in danger of getting in each other's way as we made similar points on the issue," Gore would remember. "Both of us were familiar with petty rivalries that interfered with the development of sound policy, and both of us felt so strongly about this issue that we wanted to figure out how to avoid destructive forms of competition."

Wirth would put it more wryly. "I told Gore I was not interested in running for the presidency on the strength of the climate change issue. Gore is a very competitive fellow . . . And the Senate is a funny place. People own issues . . . The question of ownership laced with politics was not good for this one."

"Ownership" settled, Wirth needed a venue for his attack. He might be a member of the powerful U.S. Senate, but a freshman couldn't even get a hearing room without permission. For that he needed the okay of his chairman on the Committee on Energy and Natural Resources, J. Bennett Johnston.

Finally Harwood went to work phoning potential witnesses. Among the scientists he contacted was a relatively "fresh face," he said, James Hansen, whose Goddard Institute for Space Studies was located on the Upper West Side of Manhattan, above Tom's Restaurant, where the famous TV sitcom *Seinfeld* would later be filmed.

"It's interesting you called," Hansen told Harwood, from seven stories above traffic spewing carbon dioxide onto Broadway. "Actually, I do have something to say."

Harwood would tell people, later, "When I heard what it was, I got excited. Tim got excited. We knew instantly this was a big deal."

———

Jim Hansen, NASA's foremost climate modeler, made his way to the witness table facing the horseshoe-shaped raised dais occupied by senators.

"I wasn't nervous. You get so little time in front of these committees."

The media had packed side tables flanking the general audience crammed—standing room only—into ten rows of leather seats. The room was impressive. The walls, marble at the base, rose in polished wooden sections separated by wooden columns set like pillars into the wood. A constellation of floodlights shone down on Hansen from the ceiling. The audience could gaze at zodiac symbols set into the walls, representing one more means, less scientific than Hansen's computers, by which humans sought to understand earthly happenings. There was the scorpion, with its dangerous stinger. The lion, leaping on hind legs.

Hansen leaned toward microphones and a trio of colored lightbulbs. The green one lit up, meaning he was free to testify.

"I would like to draw three main conclusions," he told the senators. "Number one, the earth is warmer in 1988 *than at any time in the history of instrumental measurements . . .* "

What he meant, what his written testimony would expound upon, was that the earth had warmed at a record pace for the past two decades. Four of the warmest years in the last 100 had been in the 1980s, and 1988 was expected to become the fifth. The amount of warming in the last 30 years was almost three times greater than any standard deviation, giving Hansen "ninety-nine percent confidence we can state that the warming during this time is a real trend," not a chance fluctuation.

"Number two," said the mild-mannered, dogged Hansen, "global warming is now large enough that we can ascribe with a high degree of confidence a cause-and-effect relationship to the greenhouse effect . . ."

Which meant, Arrhenius was right. And not sometime in the future. *Right now.*

"And number three," said the scientist, "our computer climate sim-ulations indicate that the greenhouse effect is already large enough to begin to affect the probability of extreme events such as summer heat waves."

Which meant that even if you couldn't say for sure that the specific, current heat wave was in any way a product of the enhanced green-house effect, in a broader statistical way, this *kind* of heat wave was.

Hansen was basing these conclusions in a large part on his NASA computer model and its ability to re-create climate, and predict it by assigning mathematical values to phenomena like rain, wind and heat. The results were tested against 100 years of records of climate reality, and judged against Hansen's years of experience studying the atmosphere of other planets.

And now, with the green witness light glowing, Hansen expounded upon predictions and told the rapt audience that greater global warming would occur over land rather than water in the future, and over polar regions rather than over the equator. Higher latitudes would grow warmer before lower ones. Warming would occur more in winter than summer.

Hansen even projected specific consequences for cities around the U.S.

With carbon dioxide doubled in the atmosphere from pre-industrial levels, Hansen said, the average number of 90-degree days in New York would jump from 15 to 48. In Washington, D.C., from 36 to 87. In Denver, from 33 to 86.

The reporters scribbled like mad.

Washington's 100-degree days could rise from 1 to 12. Omaha's from 3 to 21. Memphis's from 4 to 42.

The reporters knew they had a huge story.

In fact back in New York, later, one of the reporters, visiting Hansen's office, would bring the scientist to his window, and the men would gaze down at the "Flowers By Valli" shop and the green awning of Cafe 112 across Broadway, and the reporter would ask the scientist whether, if he was right, and carbon dioxide in the atmosphere really doubled, anything down there would look different because of it by 2030. Would average people notice any changes?

Hansen would say, "There will be more traffic on Broadway."

"Why?" the puzzled reporter would ask, thinking at first that Hansen had not heard the question right.

"Because the West Side Highway will be underwater . . . You might see Dutch engineers down there, to build dikes. The Dutch could sell their expertise in building dikes in New York, Florida, Louisiana."

"Dikes," the reporter would repeat, stunned.

"As the warming progresses," Hansen would continue, "and droughts get more severe, you might see signs in restaurants, 'Water by Request Only.' Hurricanes and thunderstorms will be more frequent. You might see tape X'd on windows across the street, against wind.

"And you'd see more police cars. Crime goes up in summers, with the heat.

"Do you think this sounds like science fiction? In a way it *is* science fiction," the scientist would tell the reporter. "But my wife and I used to go to the beach when we were first married. I'd look at the waves. Nature seems so powerful, you have to wonder. These gasses people are sending into the atmosphere. Can they *really* compete with the powerful forces of nature? Even as a scientist you wonder. But after being involved in these studies over a decade, checking the models, looking at earth's climate, figuring out what kind of force is necessary to really change nature," he would say, as if he wished the truth were otherwise, "you come to the conclusion that yeah, man can change it."

Still, it was just a theory—a computer prediction, and a rough one at that. Not to mention, as any opposing senator knew, since Tim Wirth ran his own hearing, and picked the witnesses, he'd passed up scientists who *didn't* believe that the greenhouse effect was anything to worry about. There were lots of them in 1988, charging that climate computers were so flawed that their predictions meant nothing. And now, in Dirksen room 366, senators disinclined to believe Hansen's grim predictions began grilling him the moment questioning began.

Senator James A. McLure, of Idaho, asked whether the greenhouse effect "explains the droughts of the 1930s?"

Meaning, wasn't it possible that droughts like the current one simply occurred naturally?

Hansen admitted, "We don't know what caused the 1930s to be warmer than preceding decades," and repeated that it was impossible to blame any one specific weather event conclusively on the greenhouse effect. You could talk with confidence about trends, not specific events.

Senator Wendell Ford of Kentucky told Hansen that he lived along the Ohio River, where "We have been having severe winters, and you find the ice and snow in Florida. But from the whole belt south, the winters are becoming very severe . . . explain that."

Meaning, if the world is getting so hot, smart guy, how come, where I live, it's getting colder in winters?

Hansen answered doggedly that "natural variability in the winter is much larger than in the summer . . . the first place we look to see the greenhouse signal in the summer."

Senator Frank Murkowski of Alaska suggested wryly that perhaps Congress was ready to get back into supporting the beleaguered nuclear power industry, if they wanted so badly to cut back on oil and coal.

The senators agreed that more research was needed. That's what they always said when climate came up.

But the difference today was that Hansen's assertions had captured the reporters' need for drought stories. He would make headlines across the world within hours, and vault the issue into popular imagination. Millions of people were feeling, in their guts, that there was something "wrong" with the weather by 1988. Jim Hansen had given them a reason why.

At 4:15, when the hearing broke up, the reporters surrounded Hansen, peppering him with questions.

He told them, "It's time to stop waffling so much and say that the greenhouse effect is here and affecting our climate now."

———————

Four weeks later, on July 22, a thousand miles west of Washington, a U.S. Park Service fire behavior analyst named Phil Perkins, an experienced firefighter, religious family man and a man who would soon find himself trapped in a cabin by flames, rushed from his Yellowstone Park headquarters office at Mammoth, Wyoming, hurried toward a French-built Llama helicopter and headed southwest toward the Idaho border.

He was worried about a phone call he'd just received.

Like three other fire officials in the copter, he wore a yellow button-up fire-retardant shirt, green fire-retardant pants, and he carried on his lap a white helmet and fireproof gloves. His leather lace-up boots had Vibram soles that wouldn't melt in fire.

"If I have to walk through hot ash, it won't burn my legs," he'd explain to friends about the high boots.

At 2:30 the call had come from officials in the adjacent Targhee National Forest, where woodcutters had started a blaze by dropping cigarette butts as they worked.

"The fire is small but it's moving toward the park," the caller said.

The Llama rode bumpily in the heat of the afternoon, and in hot air currents generated by a half dozen other fires already burning in the park, one of which had started minutes after Jim Hansen finished testifying in Washington. Perkins peered through a sky gray from rising ash, and marred by convection columns of smoke shaped like thunderheads, rising in all directions.

The park was so large he could not see actual flames from the fires. Yellowstone is America's biggest, oldest National Park, 3,400 square miles in size, larger than Delaware and Rhode Island combined. It is home to the largest number of wildlife in the lower 48 states, and it is so startlingly beautiful that explorer Jim Bridger's descriptions of the place in the 1840s were considered tall tales by many who heard them. Bridger described a region of forest and high meadow blanketed with swaths of sweet-smelling wildflowers: purple larkspur and pinkish fireweed and wild geraniums; with rocky ridges and steep river gorges and mile after mile of ramrod-tall lodgepole pine. Smaller, thick, twisted white bark pine clawed their way into the sun in the high winds nearer the snowline.

Bridger claimed he'd marveled at sights of great hot springs and geysers, grizzly and black bear, wolf packs, soaring eagles and whole weather systems moving simultaneously along the Rocky Mountains and Continental Divide; jewel blue skies in one place, jagged streaks of lightning in another, swirling thunderheads pummeling the park with hail or even snow, at upper altitudes, in July.

Phil Perkins considered himself lucky to have a job in this jewel of a park. It was his third year in Yellowstone, his seventeenth firefighting season, and his job as assistant fire-managing officer gave him hands-on experience at fire prevention planning, command of troops fighting blazes, responsibility for monitoring fires and search and rescue.

Fellow workers would remember him as a cheery, athletic man in his thirties who liked to take his wife and three daughters fishing, hiking and picnicking on Saturdays, and who spent Sunday mornings teaching Bible class in Livingstone, 60 miles from Mammoth, where emergency park personnel live. He was a National Public Radio junkie and a supporter of George Bush in the upcoming presidential elections, appreciating Bush's "conservative values." A compact man, mustached and bald on top, he'd put himself through Colorado State University with money earned as a smoke jumper. He'd fought forest fires in New Mexico and Alaska.

Some day, he wanted to "run my own program, move up the ladder, make more money."

Heading southwest in the copter that day, Perkins remained confident that his beloved Park Service had enough manpower and equipment to handle any fire in the park—as it had in the past. He believed fire was part of the natural ecosystem. In fact park policy since 1972 had mandated that fires that began naturally be monitored but allowed to burn unless they threatened human life or property. Natural fires helped the forest evolve.

Besides, "most fires put themselves out," Perkins liked to say.

Firefighting was considered a science, and experts like Perkins and his fire behavior analyst acquaintance Don Despain used *fire* models—mathematical representations of fire behavior to incorporate "normal" fire intensity, spread rate, wind speed, fuel load (living and dead vegetation) and fuel moisture content in making predictions about how fires would perform.

Between 1972 and 1987, the Yellowstone program had worked well. The park had recorded 235 fires, and 34,000 burned acres, 10,000 of which had been lost to natural fires in 1979, and 20,000 in 1981. Over those years, each time a fire had been spotted, Despain had been notified and, if it was a natural fire, "I'd go right out in the helicopter" to examine it, walk its perimeter, study it, incorporate its lessons in his mind.

But by July 22, after three years of drought, 1988 was the driest year on record since record keeping had begun in the park over 100 years before. The moisture content of the twigs and dry herbs on the ground—natural kindling for a fire—was as low as 2 percent, less than it would be in a stick of wood left out in Death Valley. "These dry fuels," Despain would write, "combined with high temperatures and extraordinary winds produced a series of high-pressure systems, creating some of the most severe burning conditions observed in this century."

Still, asked by a *Billings Gazette* reporter early in the season how many acres might burn that year, Despain had said, "forty thousand at most. The probability of the weather staying dry is just not very high."

It wasn't that Despain, Perkins and other "fire gods," as experts were nicknamed in the hierarchy, were bad at their jobs. It was that, like other emergency planners, medical personnel and even political leaders who would find themselves confronting "odd" or "unprecedented" weather around the world over the next decade as the planet warmed, the well-trained experts would fail to anticipate extra strains that the new extreme weather would put on them.

"We were already stretching firefighting capabilities on the day of that copter ride," Perkins would say later. In fact, the fires had spread so much that the Park Service had been ordered to try to suppress all of them.

And now, as the copter reached the border between the park and the Targhee National Forest, the four men inside gazed down through 5,000 smoky feet at flames and experienced one of those silent moments of realization that occur in history, when experts come face to face with their own vast miscalculations.

"We looked down and we knew we were off the charts," Perkins would tell friends later.

The fire, less than an hour old, had only just crossed into the park, which was obvious from the air by the sudden termination of roads down there as they hit wilderness.

"The fire was leaving a forest that was logged and entering a thick, dense forest that had never been logged. A lot of dead trees were on the ground. There was no access point if we wanted to stop the fire, which was spreading fast along the head, and spotting along the flanks."

"Spotting" meant that winds were sweeping burning embers ahead, starting new blazes.

"It was heading toward Old Faithful and maybe the town of Yellowstone," Perkins would say later. "There was nothing in its path to stop it. There was nothing in the forecast that indicated the weather would change. Hottest on record. Driest on record. Windiest."

But the moment of realization passed and the professionals got down to business. Spreading topographical maps on their knees, they flew the fire's perimeter, studying exactly what kind of dried-out superfuel lay in the beast's path. Planning strategy, they knew that forest firefighting generally involved "anchoring" a headquarters point behind a fire, so flames wouldn't come back at you. Then firefighters would try to spread around the flanks, cutting barriers between burning and unburned forest, racing the fire, hoping to encircle it in front.

Since Yellowstone authorities had forbidden the use of bulldozers in the park as permanently harmful to the landscape (as opposed to fire that was considered natural), the assault here would be made on foot. Firefighters would march in with chain saws and Pulaskis (a kind of shovel) and try to clear the edges of the fire. Airplanes would drop flame retardants along the edges.

By 7:00 P.M. that night, the fire had grown to 460 acres.

Two days later, the "North Fork Fire," as it was named, encompassed 2,500 acres and was still growing. Three days after that, a thousand firefighters were on the scene, battling the by-now 9,700-acre fire that wasn't even the biggest one in the park. The "Clover-Mist" fire alone, to the northwest, reached 46,825 acres that day, as Perkins and combined firefighting commands, volunteers and military personnel also battled the "Fan" fire, "Snake River Complex" fire, "Huck" fire and "Mink Creek" fire.

The drought continued. The blazes grew bigger.

As August began, one firefighter wrote in his diary, "Got our collective butts kicked. A lesson in humility."

Federal officials kept visiting, calling the whole thing a disaster, and Perkins or Despain, doing regular aerial surveys, found themselves looking down on scenes resembling Hieronymus Bosch paintings of Hell: lines of tiny yellow figures trudging beside enormous swaths of flame or black smoke. Fire fronts extended over 70 miles. Smoke was so thick that even in Perkins's office, miles from any fire, he couldn't see clearly from one side of the room to the other. Towns abutting Yellowstone were filled with the smell of charcoal burning, and with people evacuating.

Surely it must rain soon, thought Perkins, watching the skies when he jogged each morning, following the news on TV, driving his wife crazy with his obsession with weather.

It didn't rain. The unusually strong winds picked up.

Then the seven major Yellowstone fires began acting in ways that "fire models" did not predict. They flared up at night, instead of dying down. They shifted directions as they modified their own behavior. Fire scouts reported whirlwinds from the fires.

By Saturday, August 20, "Black Saturday" as it would be called, weather satellites confirmed the approach of more extreme winds. Dawn found Perkins commanding thousands of firefighters on the ground at the Clover-Mist fire, and staring in horror at a gigantic fire convection column, rising 30,000 feet in a thunderhead shape, from a different fire, 15 miles away at Hell Creek.

"Convection columns weren't supposed to act like that in the morning. It meant the fire was burning as hot as it does in the middle of the day. I'd never seen anything like that in my whole fire career."

But Perkins battled off unease and got back to the job at hand, giving instructions to his firefighters and keeping the Clover-Mist fire from making a run out of the park, at nearby Cooke City.

"If the fire hooks around and gets into the Soda Butte Creek drainage area," he told exhausted firefighters clustered around his map at field headquarters, "it'll be in a natural wind tunnel and the wind could drive the flames into Cooke City."

Outside that town, some jokester had changed the "Welcome to Cooke City" sign to "Cooked City."

The wind started gusting to 40 miles an hour.

Between 9:00 A.M. and 8:00 P.M., the fire raced ten miles and reached within three miles of the lodgepole pine forest outside the town. But Perkins managed to keep the blaze from getting into Soda Butte drainage, from topping high ground called Thunderer Ridge and reaching the natural wind tunnel below. Midnight found him atop the ridge, black with soot in the glow of fires and headlamps, digging a fire line with his men with Pulaskis. Every once in a while, flying embers would land in trees, and the trees would go up in a whoosh.

Perkins kept the fire out of the drainage area, even though 160,000 more acres burned that day. At least Cooke City was saved.

Surely the worst was over now, he figured, and rain would come.

But he was wrong again.

It got worse.

And that was when Phil Perkins was sent to guard the isolated cabin in the woods. To prevent, as policy mandated, fire from damaging property in the park.

———

It was on September 7, almost six weeks after the North Fork fire had begun, that Phil Perkins learned of his new assignment.

The cabin, located in the south part of the park, was usually used by rangers working the back country, who reached it by horseback or on foot. Perkins and another ranger would replace a two-man team who had been at the cabin, monitoring a fire several miles away. Because of the way the fire had been moving, nobody expected it to reach the cabin. The rangers were there as a precaution.

"My boss figured I'd catch up on sleep. I'd been working in the office around the clock. I was exhausted."

A helicopter dropped Perkins and ranger Nick Harris at the cabin, 12 miles from the nearest highway. Then it flew away.

The cabin was pleasant. It was clean and there were bunk beds and a stove. A freshwater creek gurgled outside, near the porch. The rangers who had left had cut down any trees close to the cabin. Perkins and Harris had plenty of drinking water and canned food. They had two portable water pumps, which they could supply with water from the creek. They had hoses and a sprinkler system, shovels and aluminum fire shelters that reflected heat.

As a last resort, which no one thought would happen, they could climb into the shelters and hope the fire swept over them.

"In a way, the assignment was a vacation."

Perkins and Harris went to sleep and woke up around 2:00 A.M. Looking out the window, they saw that the fire glow was a lot closer than it should have been.

"We paced from the cabin to the fire. It had gained half the distance."

They went back to sleep, and at six, got up, checked the pumps in the creek and set up the sprinkler system on the roof. They scattered hoses around the property so they would be able, instantly, to douse spot fires.

They cut up extra aluminum fire shelters and with large staple guns affixed the fireproof material to the outside of the cabin, to reflect heat. But the material flapped in the rising winds, leaving open flammable spaces.

"You can hear fire making a run at you if you're close enough. It sounds like a jet engine."

Perkins heard a sound like a jet engine.

As the flames came on, the two rangers started a defensive "backfire," using flares to light the area between the cabin and oncoming blaze. "Backfires" consume fuel that a bigger main fire would otherwise use, thereby creating a buffer zone between the main blaze and the area firefighters want to protect. Since the bigger fire draws air *into* it, it sucks in the small fire, keeping it from spreading back toward the firefighters who set it.

Perkins and Harris started hosing down the cabin and the meadow, watching the backfire.

The main fire came on, and reached within 40 yards of them. Burning embers landed on the wet roof. But the strategy worked, and the fire fell back, at least overnight.

Next day it came at them *again*.

This time it attacked from the northeast, as if it had consciously circled around. Perkins and Harris set another backfire, heard the roar when the two fires met, and saw the flames from both fires moving back and forth, "as if they were battling."

The fire fell back again.

On the third day, unbelievably, it headed toward the cabin from *another* direction.

"We had to keep hustling, wetting down the grass and cabin. It was so hot that they would dry almost right away."

Again they turned away the fire.

But by now, Perkins and Harris had barely slept. They had not radioed headquarters. "There was nothing anyone could do about it by that point and they were busy anyway." They alternated guard duty, making sure that embers didn't start a fire on the roof, that the hoses worked, that burned areas close to the cabin didn't flare up.

As dawn approached on September 10, unknown to Perkins, his family was being evacuated from park headquarters back at Mammoth. The North Fork fire, the little 460-acre blaze he had flown over a

month and a half earlier, had now burned 260,000 acres, and had traveled 40 miles.

It was probably fortunate that, as the day began, he never got to see
"Fire Behavior Evaluation No. 3," circulating among firefighters. It
read, "The dramatic behavior of September 7 and 9 is only a prelude to
conditions that are getting in place to produce a whole new level of
magnitude of fire behavior today. In fact, the system will equal or
exceed the big blow of Saturday, August 20th."

As Perkins dozed fitfully, a caravan of cars containing Mammoth residents headed out the only safe road left for them in the park, beneath a
glow on nearby ridges, as trees burst into flame and firefighters readied
for a last-ditch fight to save headquarters.

Phil Perkins woke in his cabin, looked outside and realized the fire
had circled around overnight again. It was going to make a run at him
from the last unburned direction.

He and Harris dragged themselves outside.

They set up their last-ditch aluminum pup tent fire shelters on the
grass. They heard the jet roar signifying the newest fire attack. They
doused embers and spot fires for hours, but by 7:00 P.M. the blaze still
threatened them.

At 8:00 P.M., it seemed to begin to wane.

At ten, Perkins believed they had finally beaten the monster back,
closed off the last direct path to the cabin. He had no more strength left.
The men went inside, opened cans of food and wolfed the contents
down, and then fell asleep with their boots on.

But some time that night, Perkins groaned and opened his eyes. He
was hearing the same scratchy sound on the roof that had been keeping
him awake for days—the sound of pinecones and pieces of burning
wood showering down on them.

He rushed out.

He looked at the roof.

It wasn't embers he'd heard up there.

"It was raining. We started dancing around."

———

Perkins was saved and so was Mammoth, which had been minutes away
from incineration when the precipitation had begun in that part of the

park. But the fires would continue to burn into October, when winter snows, not human effort, finally extinguished them. In the end, 1,405,775 acres would burn in Yellowstone and adjoining forests, a far cry from the 40,000 predicted by Don Despain early in the season. Despite the toll, park officials like Perkins would insist to critics that their decisions had been right, and even bulldozers couldn't have stopped the blazes.

The drought had been too great for *any* human effort to have contained the fires, they'd say. The winds had been too high, too unusual. The fuel load in the forest, having built up for hundreds of years, and having been protected by the old park policy of dousing *every* fire, had combined with the unprecedented weather to turn the park into a tinderbox.

Eventually the burned areas would start regenerating. Phil Perkins would get his wish, be promoted to chief fire control officer, and run his own program. Don Despain would become a teacher at Montana State University and spend years studying the results of the fires.

At first he would conclude that the 1988 fires should be considered "natural" because huge fires such as 1988's happened every 250 or 300 years, after the fuel load built up. The last one had occurred in Yellowstone around 1735, according to tree records, long before the industrial age began, and long before humans began putting significant amounts of extra carbon dioxide into the atmosphere.

But years later, Despain would be asked if climate change had anything to do with the 1988 fire. Sitting in his office in Bozeman, he would say, "Oh, climate is involved, all right. Acreage burned each year in the West has taken a real jump since about 1980. High winds seem to be coming more frequently."

In fact, the big, gentle researcher would warn that, in his opinion, as the new millennium began, larger fires would threaten "interface" areas throughout the West, places where cities met the wild. He would recommend that firefighters prepare for a new terrifying kind of urban fire.

He would even speculate that by studying the size and frequency of fires in the West, researchers might find a "canary in the climate mine," proof that climate was changing.

"Fire regime, frequency and size all relate to climate," he would say.

It was just a theory.

If the summer of 1988 was the summer of fire for Phil Perkins, for Bill Fraser, it was a summer of ice.

About the time Perkins was flying over Yellowstone Park's fires and Jim Hansen was testifying in Washington, Fraser was in Antarctica, about to experience an enormous flash of inspiration. He was aboard a National Science Foundation research ship named the *Polar Duke*.

Fraser wasn't a climate scientist. He didn't think about the greenhouse effect one way or the other. It was entirely outside his personal area of expertise, which was biology.

"Global warming was not even on the radar screen," he says.

Fraser, a tall, blond, earnest 37-year-old, had been fortunate enough to get a grant to work at one of the most coveted locations in the U.S. Antarctic program, Palmer Station, at the base of the Antarctic peninsula.

Antarctica, in the popular mind, might be remote and forbidding, but the truth was, life at Palmer was loads of fun. The two big prefab buildings—a dorm and lab/garage building—which sat nearby a glacier, were well heated, with a pub, a pool table and even a hot tub. Researchers enjoyed cross-country skiing on the glacier. The meals included steak and lobster and excellent pastries. The scientists were friends, the support staff cooperative. And unlike the main U.S. base, McMurdo, on the far side of the continent, there was no military presence regimenting life.

Hardworking scientists at the base studied mainly biology, with a little geology and glaciology thrown in. One team looked at krill, small shrimplike animals that formed the base of the Antarctic food chain. Another group studied ice fish, which have no red blood cells and contain natural antifreeze in their blood.

The Antarctic was a vast natural laboratory and Fraser loved it. And what a place to live! The view was so magnificent it seemed unreal. To the south of Palmer Station, where snowcapped peaks of the continent rose, the light was often lavender. In the harbor and sea, where the water sparkled it often appeared as fine gold. In the north, where 80-foot-high cliffs of ice formed a cove, on clear days, it tended to go deep blue at midday, jade at sunset. Seals lay on ice floes offshore. Penguins waddled near the base. Up on the cloudy glacier, where snow blew, the light was sometimes so smoky that it seemed the ice was on fire, merging the line between earth and sky.

But on the day Fraser had his big idea, it was Antarctic winter, dark.

"We got some twilight from eleven A.M. till one P.M. every day," Fraser said. "That was all the natural light we got."

He was in the Weddell Sea, working on the Antarctic Marine Living Resources Program, an early effort to understand how sea ice affected the continent's ecosystem.

It was still dark outside but soon it would be twilight, and Fraser had gone up to the bridge of the 219-foot-long, double-hulled ship, pondering a problem that had been dogging him for weeks.

The problem had to do with penguins, not climate.

Fraser studied two species of penguins, both of which abounded in the Palmer area. Adélie penguins were small, two feet tall maximum, white in front, black in back. Chinstrap penguins were identical looking except for a small black band, like a strap, crossing each animal's throat.

Other than the strap, the feeding habits of the two species were identical, as were their breeding habits, their summer habitats and their enemies.

So how come, Fraser wondered, the chinstrap population was rising, but the Adélie population was shrinking?

The question was really bothering him. It didn't make sense. There had to be a factor he was overlooking. Day after day, from the decks of the *Duke*, looking out at penguins swimming like dolphins, popping up and down, or weighing them in the labs on board, he tried to find a difference, *any* difference, between the two species, while the *Duke* bulled its way through the ice.

At night, lying in his bunk, thinking about penguins, he heard small ice floes sounding like chains dragging across the hull. Bigger ones sounded like Volkswagens crumbling.

"We'd been cruising in the open sea where we'd seen lots of chinstraps. We'd reached an area of intense pack ice and come to a standstill the evening before."

Now, in the seconds before the big realization hit, up on the bridge, Fraser looked out as vague light seeped into the bottom of the world.

"Holy shit," he said.

Ahead, on the ice, was "one of the most amazing sights I've ever seen." He was looking out at thousands upon thousands of penguins. He'd never seen so many in one place before. Many, still in their night-time huddles, were preening. Others, in long lines, were waddling in single file toward gaps in the ice, to feed. Thousands more had been drifted over with snow as they slept, and were rousing themselves.

The crew crowded the window to watch, awed.

Outside, wind howled through the ship's antennae.

They're all Adélies, Fraser thought.

"It was one of those lightbulb moments."

It was one of those instants when everything you know about a sub-ject, everything you've learned over years, coalesces in your mind and focuses on a problem that has, up until that second, seemed madden-ingly insurmountable.

"I saw what was different between the two species was their *winter* behavior. The chinstraps preferred mostly open water during the winter. The Adélies preferred ice."

If the chinstrap population was surging, Fraser now wondered, did that mean there was *more* open water in the Antarctic than there used to be?

If the Adélie population was plummeting, did that mean *sea ice cover was shrinking in the Antarctic?*

It was only a theory though.

Only a theory.

But in late 1988, in a hundred laboratories, thousand-year-old forests and Alpine glacial regions, in the pages of satellite reports, and even in naval submarine ice thickness records, the search was about to intensify for a human fingerprint in global warming.

Millions of dollars would be committed to learning answers. Had conditions causing the Yellowstone fire been encouraged, in any way at all, by coal-fired energy plants belching carbon dioxide into the air from Wyoming to China? To many people, such a linkage seemed inconceiv-able. Was it possible that the population of an Antarctic bird was actually dropping because people drove too many cars in Bombay and New York and Paris?

Trillions of dollars were at stake. Oil companies were waking up to the fact that they had a new fight on their hands. Tim Wirth was already drawing up legislation. And hundreds of miles south of Washington, in the mountains of Asheville, North Carolina, a greenhouse skeptic named Tom Karl picked up his ringing phone and learned the government was sending him to Europe to help investigate whether Jim Hansen's predictions were possible, whether Arrhenius's theory was becoming awful fact.

London ... Washington

T HERE HAD NEVER BEEN an effort like it. In ones and twos, in Aeroflot jets from Moscow, Lufthansa planes from Hamburg, Boeing 747s following the jet stream over the Atlantic, scientists from over a dozen countries converged on London. Each had their own belief over whether humans had begun disrupting the earth's climate or not.

Arguments started at the luggage racks. People would be shouting at each other by tonight.

Tom Karl headed past customs and into the arrival lounge, where a scientist from the British Meteorological Office was waiting for him. The American had slept fitfully during the night flight from North Carolina.

"We'll be driving to Broadway," David Parker told Karl. The village, in the English countryside, was the British government's choice as host for the upcoming meeting. Karl squeezed into the backseat of a tiny sedan and he and three other scientists—two from Britain, one 300-pound climatologist from Russia—headed west on the M40, past Oxford, on a foggy, 40-degree December day.

Dave Parker got lost.

"We were supposed to be trying to figure out global climate and we couldn't even figure out how to get to the meeting," Karl said.

Squeezed in back, watching warily as the vehicle skirted a cliff, the

37-year-old meteorologist and representative of the U.S. National Climatic Data Center was a self-described workaholic, a weather data numbers cruncher, and a lean, quiet man who could look pasty and bespectacled when in his brown sports jacket and glasses, or fit and handsome when he dressed in more tight-fitting clothes and contact lenses. Karl watched his weight, ate healthily and liked an occasional round of North Carolina golf, when he wasn't working, that is.

"There was pressure on us because of Jim Hansen's statement and other studies. The governments all wanted to know, is there *really* anything to worry about?"

In an unprecedented worldwide response, the United Nations had established the Intergovernmental Panel on Climate Change. The IPCC would access whether climate was actually changing, and *if* it was, by how much, and what the consequences might be.

Tom Karl would be one of three lead authors on the crucial chapter looking at observed climate variability and change.

"If a scientific consensus was achieved, governments would *have* to accept it because they controlled the process. They couldn't reject results," said Michael Oppenheimer, an atmospheric physicist at the Environmental Defense Fund, who had helped to set the process up—but who was not on the panel.

At the moment, though, results were very much in the future. Tom Karl "doubted whether the science would even be able to show a link. I wondered, with so many scientists coming, how we could say anything we could agree upon, and how we could make any sense at all."

At the meeting, scheduled to last three days, roughly 40 participants would present their own work, find gaps in research and come up with a strategy by which they could produce their report by 1990. Two more studies would be due in 1995 and 2000, and all three would be reviewed by thousands of scientists and government officials in every country, before being revised and published.

Karl had no idea that soon, because of the reports, careers would be under assault. Families would split up. Politicians in high offices would try to discredit researchers, and diplomats who'd never studied climate in their life would battle to change wording in the reports.

At the moment the process seemed purely scientific.

"There was an urgency to it. We all had a sense that we were participating in something greater that any one person. We would be forced into trying to make sense of a broader picture, something that never would have happened otherwise."

In late afternoon Karl's little car finally turned onto a forested two-lane blacktop bringing them to Dormy House, on a hill overlooking the charming village of Broadway. He gaped as they pulled up to a converted seventeenth-century farmhouse, with sprawling manicured lawns, ivy-covered walls and a shrub-bordered parking area filled with Mercedes, BMWs and Jaguars, vehicles the inn's usual clientele could afford. It was suddenly clear to Karl, who normally slept in cheaper government-expense-account-lodgings, that he had arrived at the moment when greenhouse theory entered the realm of international policy. The sumptuous inn gave Karl a pretty fair idea of all the money, power and attention his little area of expertise would receive for the rest of the century.

And by the end of the century, he would figure prominently in the fight.

His involvement was an outgrowth of a fascination with weather that had started by age four, back in the Chicago suburb of Niles.

"We had a weather vane, barometers and anemometers. I loved to watch the sky."

By five, he was fighting over climate.

"My dad and I would each write down a forecast for the next five days. We'd include precipitation, and high and low temperatures. The subjective part was clouds. He might predict 'partly cloudy.' I'd have 'cloudy.' We'd argue over who was right.

"I was thrilled by extreme weather, especially snowstorms. I used to wait for winter. A good snowstorm," he'd recall years later, grinning with pleasure, "would be *great*. I wouldn't sleep. I'd keep my face pressed to the window. I'd be watching for the way, when the wind shifted, the flakes would start blowing from a different direction. Or the minute when," he'd say, rapt as a sports car lover describing the newest Porsche, "snow would change to rain."

Later, at Northern Illinois University, he skipped humanities classes and enrolled in a meteorology program, "the best thing that ever happened to me." Then he entered the University of Wisconsin masters degree program in Meteorology and got a part-time job as a radio weatherman on WKOW, an ABC affiliate.

It was here that he was first embarrassed in public when the weather didn't do what he thought it would, and he never forgot the lesson he would bring to Dormy House.

"It was a Sunday, and I'd been cross-country skiing. I headed into the station to tape-record a forecast that would be broadcast forty-five minutes later . . . That night was partly cloudy but otherwise clear, and the last weather map I'd seen, on Friday, had said chance of flurries but not much. But when I got into the station I couldn't believe it! Heavy snow was breaking out in western Illinois and southern Iowa. I checked model simulations of the storm and it was moving much farther to the south on them."

The weather models, and not for the first time in Karl's experience, had been wrong.

Alone in the station, unsure what to predict, young Tom Karl found himself ripping sheet after sheet off the weather teletypes, trying to figure out what to tell his audience.

"By nine it had started to snow outside. At nine-fifteen I decided to contradict the models and predict four to seven inches for Madison, and one to two inches for La Crosse, further north.

"Well, that storm just wracked up unbelievable numbers. Madison got five inches, and people said to me, next day, great forecast! Except I'd been right for the wrong reasons. It only stayed at five inches because the snow turned to freezing rain.

"But in La Crosse, they got fourteen inches! I was a hero in Madison, but not up north.

"That storm formed part of who I am. I never forgot, you can *never be complacent about the weather*. Same thing with the greenhouse effect. *Never take anything for granted*. The one thing you can be sure of is, there will always be a surprise in weather, and you hope it won't be a major embarrassment."

By the early eighties, he'd given up forecasting and come to work at Asheville as "a plain old scientist," concentrating on his two great professional passions, weather extremes and historical records. That his boss told him the job held virtually no opportunity for advancement did not bother Karl. He was doing what he loved.

"We looked into a 1980 heat wave and drought and tried to put them in perspective. Same thing for cold snaps in 1981, '82 and '83.

Instead of just looking at the events, we said, how can we look at this in terms of climate?"

And now, in England, as he checked into Dormy House, he was to try to put the whole *earth's* climate in perspective, only instead of the audience being just Wisconsin, it was the entire world. Since Hansen's testimony in Washington, months earlier, the climate issue had grown in prominence. George Bush, newly elected president of the U.S., had declared, "Those who think we are powerless to do anything about the greenhouse effect forget about the White House effect. As president, I intend to do something about it."

He hadn't done anything yet, but he had appointed William Reilly, former head of the World Wildlife Federation and a firm believer that CO_2 emissions were changing climate, as new head of the EPA. Environmentalists hoped the appointment meant Bush was serious about tackling the climate issue.

Tim Wirth had been more active. After Hansen's appearance, he'd given a major speech at a Toronto conference calling for international cuts in the use of fossil fuels causing carbon dioxide emissions. And by the time Karl sauntered into Dormy House, Wirth was pushing a National Energy Policy Act, to "establish a national energy policy that will reduce generation of carbon dioxide and trace gases as quickly as feasible."

Major provisions of the bill directed the Secretary of Energy to draw up a "least-cost national energy plan," ordered studies on ways to replace traditional coal-use technologies with cleaner ones, supported nuclear power and directed the Congressional Budget Office to study the possibility of instituting a national CO_2 emissions tax.

Although the bill had died without being considered by the 1988 Congress, Wirth planned to reintroduce it in 1989.

Tom Karl liked the idea of more research but considered it premature to make major policy decisions like raising taxes before more scientific results were in.

"First, I had big questions about urban heat islands and the way they affect conclusions about global heat."

What he meant was, temperature records often come from urban areas. But towns and cities generate their own heat, from cars, factories and heat-absorbing tar and concrete. So were temperature records *really* reflective of rising *global* temperatures, Karl wondered, or just urban heat islands?

The arguments grew severe at Dormy House.

"One person would say 'the urban-heat-island effect totally negates any warming on record!' Someone else would say, 'No! I've done a study and if you remove the big cities you *still* have strong warming!' "

Karl also worried about the validity of ocean temperature data that showed the seas warming.

"In Asheville we'd been working on putting together an ocean temperature data set with colleagues in Boulder and the UK. But we were aware there might be large biases in the data. The ocean temperatures had been measured over time, but the *way* they'd been measured had *changed* over time."

So again, even though records showed temperatures rising, had they *really* risen? Or had the way they'd been measured masked opposite conclusions?

For instance, back in 1888 sailors had taken ocean temperatures "by using millions of samples from wooden buckets. They'd drag up water and take measurements. Later they changed to canvas buckets, and then to readings from ship intake valves. And more recently, temperature is measured with contact surfaces outside the ship. All these different ways of measuring had different implications. What thermometers were used? How long did the water sit on deck before readings were taken? Was it cloudy or clear when they were making measurements?"

Of course, to the griping sailors assigned the laborious job of hauling up buckets in the North Sea, at a time when moon walks were science fiction, and carbon-dioxide-spewing Ford Explorers were not even yet a dream in some inventor's mind, it would have seemed inconceivable that, a century later, the great scientific minds on the planet would be fighting over water temperatures in the buckets, or how to impose consistency onto land readings taken in 1900 by German sergeants in colonial Dar es Salaam, British cattle ranchers outside Nairobi and Spanish sugar barons near Havana.

"We had some real arguments," Karl recalls.

And if Dormy House wasn't stormy over the temperature records, Tom Karl could walk into the meetings that went on all day, or in the pub at night, and hear the arguments over climate models like Jim Hansen's.

"There were big questions over whether models could adequately reproduce the climate system," said Karl, who knew the models were supposed to predict climate by assigning mathematical values to different influences on it, but some key influences weren't even included at all.

"The models were simplified," Karl says. "They didn't include the way oceans affect atmosphere. They didn't include other man-made factors beside carbon dioxide that force climate, like sulfates (which cool the atmosphere). The models were coming up with warming rates that seemed high relative to what temperatures were really doing."

In short, "There were lots of theories, but evaluation was far behind."

Karl's last big question was, even if scientists agreed that earth's temperatures were rising, how could they be sure that a heightened greenhouse effect was responsible?

Maybe natural climate variation explained the rise.

Dormy House was a tower of scientific Babel.

"Everyone had their own ideas, and remember, scientists get recognition and satisfaction by coming up with their own innovations of how science works. So everyone came to the meeting wanting to show that the work they're doing was the best representation of the real climate, or best hypothesis explaining why things were or were not changing. It was in no one's interest to say they believed anyone else. Science works by trying to jettison your colleague's ideas so things make better sense, until you get to a point that ideas are so well formulated that no one can toss 'em out."

By the end of the three days Karl didn't know any answers, but at least he knew how the questions would be attacked over the next two years.

"People had decided to work with each other, something that never would have happened without the IPCC process. If there was an issue close to being resolved, scientists said, 'we might go further in that area if we work together.' "

The research would not be designed to reach any particular conclusions. The goal was to get answers so valid that governments could act on them. Tom Karl had no interest in being proven wrong in front of the whole world.

Meanwhile, extending outward from the little village of Broadway, reaching decision makers in thousands of labs and universities and research programs around the world, were the key questions that needed to be solved about global warming. Government and private research programs would now begin to target questions like Karl's. And those questions represented only part of the IPCC process. Other chapters would gauge the consequences of climate change, to hopefully enable policymakers to make informed decisions.

Tom Karl and other lead authors would now go home, read all the scientific papers related to their chapters, assess the state of knowledge on their subjects and continue their own work. Their chapters would pass along what they'd learn.

When Karl reached Asheville his wife asked how the meeting had gone.

He spent too much time at the office as it was, she thought, and she did not like his answer.

"It'll be a lot of work."

———

Months later, on April 26, 1989, a 63-year-old former chemist who would make environmental history that day rose early in her London attic flat, cooked an egg for her husband in her tiny kitchen, drank her daily cup of coffee and made her way downstairs, expecting prominent visitors.

Shortly after nine, on foot or by taxi, half a dozen British Cabinet ministers and 30 climate scientists began arriving at the small cul-de-sac where the former chemist lived.

The invitations they carried read, "You will know of the concern that the accumulation of so-called greenhouse gasses in the atmosphere could result in a significant man-made change in the world's climate.

"The Prime Minister would therefore like to hold a seminar to enable her and her ministerial colleagues to hear at firsthand the opinions and advice of some of the foremost experts on climate change and to discuss this with them informally."

One of the only two Americans to receive the invitation, Michael Oppenheimer, had even phoned the British UN ambassador, Sir Crispin Tickell, to make sure the understated note wasn't a joke.

Politically far-left liberals like Oppenheimer had never associated today's host, the former chemist, also known as "Attila the Hun," "The Iron Lady," or Margaret Thatcher, prime minister of Great Britain, with environmental causes. Oppenheimer was more familiar with her ten-year record of Ronald Reagan–style supply-side economics, her tax cuts at the expense of public spending, her deregulation of British industry and her savage opposition to trade unions.

American conservatives revered Margaret Thatcher. Michael Oppenheimer did not. And so far, in the U.S., the climate fight was breaking down along traditional liberal–conservative lines. Liberals, who tended to distrust large energy companies, tended to believe claims that oil and gas use was damaging the atmosphere. Pro-business conservatives were more likely to consider the science unproven, or junk. Moderates didn't know what to think.

Anyway, Crispin Tickell had assured Oppenheimer that the invitation was no joke, and Tickell should know, since he was the one who had convinced the prime minister to hold the meeting, who had helped convince her the greenhouse effect was threatening a stable climate of earth.

Ten Downing Street is actually three adjoining brownstones, blocked off from Parliament Street by an iron gate guarded by bobbies. The brownstones were built in 1683 by Sir George Downing, Oliver Cromwell's spymaster, as a speculative development. They've been home to British prime ministers since 1732, when Sir Robert Walpole accepted the complex from King George II and discovered that Downing had built the house the same way he conducted politics—shabbily. Repairs to 10 Downing Street have drained the British Treasury for almost 300 years.

Margaret Thatcher had done some remodeling and stocked the place with silver pieces and china borrowed from the British Museum, as well as borrowed portraits of her favorite achievers, like Lord Nelson and Sir Isaac Newton.

The museum "hides their best stuff when they see me coming," she liked to joke.

But today's meeting would be no joke. The British Meteorological Office and University of East Anglia's Climatic Research Unit had just announced that the six warmest years on record on earth had been, in order, 1988, 1987, 1983, 1981, 1980 and 1986.

By 9:30 Thatcher was ready, and it was a proud moment for Tickell, a professional diplomat and dogged aristocrat, who had first become interested in climate change while on sabbatical at Harvard, from the British Diplomatic Service, in 1975.

"What stimulated me was a CIA report I'd read on the effect of weather on politics. It made me think climate would be worth studying. But at Harvard, the director of the Center for International Affairs thought it was not a foreign policy subject.

"I told him, yes it is."

Harvard and the Diplomatic Service both gave in, in the end. "I'm an eccentric, so they figured I might as well be indulged," said Tickell, who, after studying the issue, had grown more concerned about it. He'd had a chance to discuss it with Margaret Thatcher in 1984, in a plane, coming back from a Paris summit meeting of the leaders of big industrial nations. Tickell represented the European Commission in the group at the time.

"I infiltrated her mind," Tickell wryly recalled.

By April 1989, the prime minister had been converted. Today's seminar was designed to educate her ministers and set the stage for an announcement she planned to make in the afternoon, to reporters already gathering outside.

"As we went into the meeting, she looked at the ministers and said, 'You're not here to talk. You're here to listen,'" Tickell said.

Oppenheimer remembered, "The ministers looked scared shitless of her."

All through the morning the scientists gave presentations, Thatcher asked questions and the mightiest ministers in England shut up.

"One minister fell asleep. The next week he was fired. I think it was a coincidence but I'm not sure," Oppenheimer said.

As the hours ticked past, researchers from the British Meteorological Office basically repeated the concerns that Jim Hansen had voiced in Washington. Their own work mirrored his findings, and their worst-case scenario predictions of heat rise, storm increase and global destabilization paralleled ones being made in the U.S.

Thatcher joked, "If we turn off one of these big chandeliers we'll get a big jump on energy conservation." She told the scientists she had installed energy-efficient lightbulbs upstairs, in her living quarters, but they were not providing enough light.

As the group broke for an informal lunch, Oppenheimer went up to Thatcher and told her he was impressed that she was spending "so much time with a bunch of scientists."

"Oh, people think we're so stuffy," she replied.

"She didn't take a break for any other government business," Oppenheimer recalled. Even more astounding to the political lefty were her remarks that free-market policies, which she usually favored, would not solve the climate problem.

After lunch the presentations resumed, the ministers seated politely behind Thatcher, the scientists seated in front. She told them that while there was a need for more research, governments could not afford to do nothing.

Six and a half hours after the "teach-in" began, Thatcher led the scientists outside, where dozens of reporters began shouting questions.

The *Independent* summed up her answers on the next day's front page. "Margaret Thatcher wants to take the international lead in combatting the greenhouse effect of global warming . . . The Department of Energy's free-market policies may be reversed, with more government intervention to reduce the growing consumption of coal, oil and gas."

Thatcher had just become the first major world leader to call for global action against the use of fossil fuels.

Cynics charged that her whole purpose was to promote nuclear power. They claimed that by 1989, European environmentalists were wielding so much power that Thatcher wanted to appease them.

But Thatcher had never been big on appeasement. She had always taken pride in her background in chemistry, and she'd also been pleased with the role British scientists had played in finding the ozone hole over the Antarctic, another atmospheric problem. Thatcher had been active in promoting a global accord already in effect by 1989, to cut down on dangerous aerosols causing ozone destruction.

The truth was, although American and British conservatives saw eye to eye on many issues, climate change wasn't one of them.

"British conservatives have more of a tradition of stewardship, of wanting to hand over a better world to their offspring," said John Gummer, Thatcher's future Minister of Environment. "In America, business *is* conservatism. In Britain, conservatism is more of an outlook that may not be tied as closely to short-term business interests.

"Also, there is a kind of religious aspect to American conservatives, a belief that things will always work out. Through science. Through religion. Through whatever. British people do not have the experience of 'things working out' so easily. Sometimes you have to help."

———

Shortly after Thatcher's seminar, in Washington, Jim Hansen came face to face with the different ways the American and British governments handled the climate issue, a crucial question, since any attempt to curb international fossil fuel use would be doomed without U.S. participation.

Called to testify again on Capitol Hill, this time before Tennessee senator Al Gore's Subcommittee on Science, Technology and Space, Hansen sent a copy of his prepared remarks to the White House Office of Management and Budget for review. The office routinely checks government employee testimony to be presented to Congress to make sure it is in line with the president's budget.

But looking at the testimony after it came back, Hansen got a surprise.

"They changed it."

Hansen had written, *before* the change, "An increasing greenhouse effect undoubtedly implies some broad changes in the nature of regional climate . . . Although these results require confirmation from other models and appropriate observations, certain fundamental conclusions emerge which we believe are very unlikely to change as knowledge of the climate system becomes more precise."

In other words, in Hansen's opinion, his basic conclusions were right, but smaller details might change after further research.

But *now* the testimony read, "Again, I must stress that the rate and magnitude of drought, storm and temperature change are very sensitive to many physical processes . . . some of which are poorly represented in the models. Thus, these changes should be viewed as estimates from evolving computer models and not as reliable predictions."

In other words, in the rewritten text, Hansen's basic conclusions were very much in doubt even to him.

Hansen was no stranger to high-level disapproval. His bosses at NASA had tried to dissuade him from testifying in 1988 and had told him afterwards about calls from the Reagan White House "expressing great displeasure about my testimony."

He'd hoped things would be better with George Bush in the Oval Office, but now he was getting calls from at least one senator saying that Bush's chief of staff, John Sununu, became enraged if Hansen's name came up. Sununu didn't trust climate models and considered the whole climate scare a strategy by environmentalists to limit economic growth.

As the day of his appearance before the subcommittee approached, Hansen wasn't sure what to do. His wife had been diagnosed with cancer. He wasn't sleeping very well. "I didn't come from an intellectual background. My father was essentially a sharecropper," the scientist said.

Trying to figure out the right course of action, he found himself late one night reading *The Grapes of Wrath* and thinking about the novel's protagonist, an Okie farmer who goes to California during the Great Depression and ends up opposing authorities over issues that he thinks are right.

Hansen went to the press and to Al Gore, and he told them what had happened.

He found himself, once again, in the headlines. Gore charged elements in the White House with being "scared of the truth" and with trying to keep the administration from proposing a UN treaty to combat global warming.

White House spokesman Marlin Fitzwater, admitting that the testimony had been changed, claimed the culprit was "five levels down from the top."

Fitzwater told the press, "President Bush's personal view is that this is a serious problem . . . but the science is something that still has to be sorted out."

Hansen had now embarrassed the president. Political editorials all over the country mocked Bush and threw his "White House Effect" speech in his face. And if John Sununu was displeased at the bashing, his ire could only have grown one day later when, in New York, Crispin Tickell stood up at the UN and called for an international treaty to deal with the problem of global warming.

Margaret Thatcher reiterated the call in a speech to the General Assembly months later. She called upon the world to complete a treaty *mandating* cuts in greenhouse gasses by 1992, when the World Environmental Conference was scheduled to be held in Brazil.

Limiting emissions would be hard, she admitted, because it would involve restricting the use of fossil fuels.

But difficult or not, the process had momentum. The General Assembly began drawing up a resolution to create a negotiating body to draft a climate stabilization treaty.

In North Carolina, Tom Karl read about the speech, and about how Thatcher advised negotiators to wait until the IPCC report came out before making detailed plans.

The pressure was growing to get it right.

Then, as Karl had predicted, Mother Nature threw in a surprise.

If Tim Wirth had gleefully watched record heat waves call attention to his hearing back in June 1988, he now picked up his newspaper daily and read headlines about record cold.

On Thanksgiving, a snowstorm dropped 4.7 inches in Manhattan and broke snowfall records across the East Coast.

By December 5, the unabating cold was causing subway shutdowns in New York as equipment broke.

By December 20, ice had closed much of the Mississippi and Missouri Rivers.

"The cold is historic," John H. Lichtblau, an oil industry analyst, told the *New York Times*. "It is outside the normal cycle of marketers serving the Northeast."

If public policy is to any extent a reflection of public will, and public will is affected by press coverage, the record cold spell was extremely inconvenient for Wirth and Gore. It was tough to talk about global warming when heating bills were doubling from Florida to Maine.

In newsrooms around the country, the global-warming debate began turning on its head. Greenhouse skeptics were now the ones pointing to their thermometers. Jim Hansen's supporters were suddenly arguing that no single round of weather reflected the big picture.

In Washington, Tim Wirth opened his mail and found a political cartoon from the United Feature Syndicate which was being published throughout the country. It showed two middle-aged men in a raging snowstorm; the first holding a sign reading "BEWARE OF GLOBAL WARMING" as snow accumulated on his shivering shoulders. The aging hippie had a beard like poet Allen Ginsberg, granny glasses like John Lennon, a frayed old pea coat with a Vietnam-era peace symbol on it, and pants striped like the American flag.

The second more respectable-looking man, John Q. Public, walking past, held his hat against the wind as his eyes swiveled to the protest sign in amazement. He pressed a folded newspaper against his coat. Its headline read, "RECORD COLD."

By December 23, 125 cities around the country were reporting shattered temperature records. Snow fell in Houston, and power failures left 12,500 homes without electricity in Fort Worth. Indianapolis thermometers read 23 below zero.

Governor Bob Martinez of Florida declared a state of emergency as Miami's temperatures plunged to 31 degrees, and the wind chill factor in Orlando reached 5 degrees below zero. Florida highways were so icy that traffic was stranded, motels full, and emergency shelters open for the motorists being rescued by the National Guard from their cars. The state's $3.5 billion citrus crop and $1.6 billion sugar industry were in danger.

The cold extended far beyond the U.S. As far away as tropical Bangladesh, a freak cold wave was killing people.

Was it any coincidence that as temperatures plummeted across the country, readers of the *New York Times* sat in their heated apartments and read the headlines, echoed on TV news, "SPLIT FORECAST: DISSENT ON GLOBAL WARMING. Skeptics are Challenging Dire 'Greenhouse' Views."

The article said, "Exactly how many scientists are involved in serious climatic research is unclear, but experts in the field say it includes fewer than 300 climatologists, meteorologists, geophysicists and people in related fields. Many of them, perhaps the majority, have not taken a firm position on the debate . . ."

One skeptic, Richard S. Lindzen of MIT, held the same honorary chair that Jule Charney, chief author of the 1979 Climate Research Board report, had occupied. But Lindzen's view of climate change differed so vastly from his predecessor's that he had written a letter to President Bush, saying, "Current forecasts of global warming are so inaccurate and fraught with uncertainty as to be useless to policy makers."

Another skeptic, Patrick Michaels, former president of the American Meteorological Society, called the climate science "murky," said cutting fossil fuels would cost billions, and predicted dangerous levels of government control over the lives of average citizens if obligatory energy cuts were instituted.

"The prevention of global warming (would be) the greatest centrally planned social engineering experiment in history," wrote Michaels in an op-ed piece for the *Washington Post*.

The skeptics had been there all along. It had just taken a few months and a cold spell for them to become more visible, and now they were about to find patrons too. If Tom Karl, in London, had reached the intersection point between science and policy, Pat Michaels now picked up the ringing phone in his University of Virginia office, and the stranger on the other end, Frederick Palmer, brought him to the point where science and corporate lobbying meet.

The friendly-sounding Mr. Palmer told Michaels he worked for the "Western Fuels Association" and invited the professor to his office in Washington.

"I was wary of the guy but I went."

Michaels had opposed the idea that humans could substantially change climate long before he ever heard of Palmer. The professor was a voluble man with a quick tongue, an artful writing style and a surprisingly sensitive ego for such a combative fellow. For years it would bother him that Al Gore would not shake his hand after a Senate hearing.

Physically, Michaels bore a facial resemblance to the impish-looking actor Henry Travers, who played the angel "Clarence" in the Christmas film classic *It's a Wonderful Life*.

He took pride in the fact that as far back as grad school, he'd opposed popular climate theories that seemed wrong to him.

"People back then were talking about an ice age coming. Global cooling was the rage. I had trouble with that too."

And now, driving to Washington, Michaels was still taking umbrage at anyone who thought the earth's climate was "fragile" or easily turned by humans from its natural processes.

"The atmosphere isn't sensitive, godammit," he liked to say.

As Michaels would tell anyone who asked, the climate-panicked segment of the population was merely undergoing the same basic education that every first-year meteorological grad student experiences when they learn how humans influence the way solar heat gets reflected back to space. It happens any time a white reflective surface, like a desert, is changed into a black heat-absorbing one, like a parking lot. It happens when cities or towns go up, or if a section of ice-covered ocean melts, so the formerly white area now appears black from space.

"Every beginning grad student looks at the equation and says Wow! We're gonna change the climate in a big way! It *must be sensitive!* And the public are all amateurs. They're not professionally trained, so by definition they have that freshman response. But if the earth's really that sensitive how come we're still *here?*" Michaels liked to challenge his students. "How come the climate didn't implode a long time ago? After all, the surface temperature of the planet is only fifteen degrees above the freezing point. So you see, any apparent instability is buried in a larger stability. Otherwise earth would be a frozen ice ball."

So now Michaels drove into downtown Washington to meet Fred Palmer, a former lawyer and current president of Western Fuels Association, a coal cooperative owned by electric utility companies that burn coal for power.

"We supply the coal," Palmer liked to tell people asking what a coal cooperative did. "We have our own mines and we buy coal from other producers. We have rail cars that the railroads pull for us. The municipal utilities that own us serve electricity to a fourteen-state area in the middle of the country."

Palmer's salary, as well as the agreement he was about to sign with Michaels, was being paid for by electricity users in the Denver and Albuquerque suburbs; in Billings, Montana; Bismarck, North Dakota;

Cheyenne, Wyoming; Rochester, Minnesota; and dozens of other municipalities from the Mississippi River to the Rocky Mountains, and from the Canadian border to the Rio Grande.

And Palmer was a dapper, friendly man, bright, blue-eyed and Nordic looking, who had worried over this newest assault on his industry ever since Jim Hansen testified before Tim Wirth and *Sports Illustrated* magazine published a spread on the greenhouse effect.

On the day Hansen had testified, Palmer had strolled into his house, switched on the 6:30 news and watched Peter Jennings swim up on the screen.

"It was the whole nine yards. The apocalypse. Those hearings. Yellowstone Park was on fire. The country was in a drought. The timing was on purpose. *That's how they work.*"

Then he'd read the *Sports Illustrated* article, on the porch of his North Arlington, Virginia, home, overlooking his pretty garden.

"I thought, here they come."

"They," of course, was the environmental community that had sought for years to limit the use of coal, tax it more, or restrict coal company access to public lands. Palmer had fought them over the Endangered Species Act. He'd fought them as an attorney, when they tried to get the Supreme Court to deny the coal industry access to leasing land in Wyoming's Powder Hill Basin, under the National Environmental Policy Act, until environmental impact statements were completed.

The "enviros," as Palmer called them, were always coming at him one way or another, and now Palmer saw them coming again.

"I thought, my God. This is it, a clear and present danger to our common well-being."

His concern had grown over the following months with each new article on potential problems from the greenhouse effect.

"*Time* magazine had the greenhouse effect as a cover story. The issue could lead to restrictions in the use of fossil fuels."

Still, fighting it didn't require big mobilization right away because "the political-policy process is slow. There's huge inertia in the system . . . and besides that, Republicans controlled the White House. But pressure started in Congress and the Senate. Tim Wirth. Al Gore. It started in the press and inside the Beltway, with magazine and think pieces and more hearings."

Western Fuels had "no program in place" to do combat yet, and the big oil and auto companies wouldn't weigh in until later, but Palmer decided to start his efforts early.

"I had to come at this from the standpoint of a lawyer running a big lawsuit, wherever it's going to be. Kyoto. U.S. Congress. PBS. *Nightline* . . . You find experts who support a point of view that is compatible with your business interests."

Palmer had been reading voraciously and attending conferences to find his experts. He was making his list. Sherwood Idso was a professor in Arizona who had a theory that increased carbon dioxide was actually good for the planet because it accelerated plant growth. Richard Lindzen at MIT believed that the roles of clouds, water vapor and oceans were completely misrepresented in the models and could actually mitigate any increased heat caused by man-made gasses in the atmosphere.

Pat Michaels, invited, appeared in Palmer's office.

Michaels says, "He was obviously not a poor man. Nice office! Filled with western art, artists I didn't recognize but I thought expensive. Maybe even a Remington. Palmer had my article on his desk: 'The Greenhouse Climate of Fear.' "

Palmer told Michaels that the piece represented "the only argument I've heard that rings true on this issue," and Michaels thought, "That's nice, but I still haven't the slightest idea from his tone where this is heading."

Palmer "tried to find out if I was really knowledgeable about the issue," and Michaels had no trouble going on about his lifelong fascination for a while. Like Tom Karl, he had kept his face pressed to windows, during storms, since age five.

"Only one side of the argument gets publicized," he said.

Palmer asked Michaels if he would be willing to produce a regular climate research review, to be published and distributed to interested parties. Such a review, of course, would constitute evidence against the notion that humans were dangerously changing the climate. Palmer would help make it available to journalists, legislators and also throughout the coal industry, which needed to be "educated" about an issue about to explode in their faces, he felt.

Palmer the lawyer was planning his case.

Michaels said sure he could produce a review. The contract was drawn up. The University of Virginia got the money.

Soon after, journalists and legislators began receiving copies of a newsletter called *The World Climate Review*.

———

Despite efforts to influence Congress, in the end the climate itself would be the key factor. Even the most ardent believers in a coming age of rampant global warming, like Jeremy Leggett of Greenpeace, knew that "three years of continued calm or cold weather would hurt efforts to curb the use of fossil fuels." But as cold weather continued across the U.S., Jim Hansen's confidence in his predictions did not falter. In January of 1990, the controversial scientist found himself in front of an auditorium filled with climate researchers in Washington.

Hansen offered the room a $100 wager.

"I'll bet anyone that one of the next three years will be the hottest on record," he said.

The 90 scientists looked at each other, but only one of them took the bet: Hugh Ellsaesser, a climatologist at Lawrence Livermore National Laboratory in California.

When Hansen got home to New Jersey, his Dutch-born wife Alliek told him she disapproved of the betting. She thought it demeaned science.

"She kept talking about Pete Rose, the Cincinnati Reds baseball player. Rose had been kicked out of sports for gambling."

Hansen laughingly assured Alliek that he wouldn't make any more bets in public.

Then he and the other 90 scientists who had been in that room waited to see who would win the bet.

Great Britain ... North Carolina

EVANGELISTA TORRICELLI, ex-assistant to the scientist Galileo, presented the public with a historic device of his invention in 1643. He called it a mercury barometer and said it could measure air pressure. The "barometer" consisted of a tube, open on the bottom, standing permanently atop a container filled with mercury. When air pressure rose, Torricelli said, the air would push down on the mercury and force it into the tube. The higher the mercury rose, the higher the air pressure that caused it.

Even in 1643 scientists understood that air pressure changed when storms arrived, or as people climbed a tall mountain. Air seemed thicker close to the ground. Higher up, it grew thin and climbers became dizzy. Why this happened was beyond scientific explanation. To suggest that it might involve factors other than God's will was to risk punishment from the Inquisition.

What neither Torricelli nor Galileo could have guessed, in a time when a belief in molecules—tiny particles making up matter—might have earned a burning at the stake (Galileo was imprisoned for insisting the earth moved around the sun), was that air pressure—a starting point of wind, storms and the terrible events that would sweep across Europe on January 25, 1990—exists because the invisible atmosphere is composed of molecules. The molecules are perpetually moving and, through the constant force of their attack, creating pressure against any surface they strike, from an aching human eardrum to the wing of a Boeing 737 climbing toward its cruising altitude of 27,000 feet.

Galileo, famous for dropping objects off the Leaning Tower of Pisa to study what would one day be called gravity, never could have guessed that air pressure is greater at low altitudes because gravity pulls air molecules toward the planet, packing them into the smaller space at the surface.

Thanks to gravity, on an average day on any sea on earth, a proud sailor standing on the deck of his brand-new, 36-foot Benetau racer feels an air pressure of roughly 14.7 pounds per square inch against his bare chest.

Gravity is weaker the higher you travel from earth, and at upper altitudes, the fewer molecules of air present have more space in which to move around. Were the sailor to board a jet and open the emergency exit at 20,000 feet, he would experience, as he was sucked out into the atmosphere, air pressure of only 7.1 pounds per square inch.

By January 24, 1990, a day before the deadly storm, when British meteorologists found themselves looking at the latest disturbing photos provided by America's new NOAA-11 satellite—which also monitored global warming—lots more was known about air pressure than during Torricelli's time.

The meteorologists knew that several other factors besides gravity also affected air pressure and that all of them, at the moment, seemed to be conspiring against Europe.

They knew that when air heats up, the molecules inside it move faster, causing air pressure to rise. That's why during cold months, air pressure tends to *shrink,* causing low-pressure areas over oceans near the poles to expand. Since low-pressure areas are cloudier and rainier in general, winter storms tend to be worse than summer ones.

The meteorologists also knew that wind occurs when air from a high-pressure area pushes into one of low pressure, in an effort to equalize the two pressures. The greater the difference between the areas to start with, the more violent the winds they produce.

They knew that when wind blows over land, friction between air and the earth's forests, buildings and mountains can actually slow it. But over oceans, no friction slows winds so they tend to blow more strongly.

They knew that borders between air masses, called "fronts," tend to be where most severe weather occurs.

Finally, they knew that just as large differences in temperatures of bordering air masses can cause nasty weather, steep temperature discrepancies between neighboring sections of ocean can heighten storms too.

What they were watching on January 24 was a developing worst-case scenario, with all these conditions happening at the same time.

Today's coming storm was a byproduct of the frigid weather that had battered the U.S. while Jim Hansen made his climate change bet in Washington and Tom Karl began his work on the IPCC report in Asheville. By January 24, the cold system that had depleted stocks of heating oil in the U.S. had moved off New England, over Labrador, Canada.

Sea surface temperatures were abnormally cold east of Newfoundland, but just south of that, in the Gulf Stream, they were oddly warmer than usual.

"This enhanced temperature gradient appears to have been present over the (whole) North Atlantic," the *Journal of Meteorology* reported.

Meanwhile, a strong jet stream was pushing the cold air toward Britain, which, like most of Europe, was enjoying its fifth warmest winter since 1659.

The fury of what was about to happen contrasted with the quiet alarm pervading the second-floor central forecasting headquarters of the British Meteorology Office in Bracknell, an hour from London by train, in a tall, ugly complex topped with antennae and dominating the low skyline. The office, like climate science in general, had come a long way since the establishment of government weather monitoring had been first discussed in the House of Commons in the 1850s, and howls of laughter greeted one member's comment that some day Londoners might predict weather 24 hours in advance.

But the service had been approved and its first head had been the brilliant, psychologically tormented Captain Robert FitzRoy, former commander of the HMS *Beagle*, the ship on which Charles Darwin had traveled the world and developed his theories on evolution. FitzRoy himself did not believe in evolution. He believed that the extinction of dinosaurs had been caused by their inability to fit into Noah's Ark.

When it came to weather, though, he'd relied on science. FitzRoy had seen how barometers at sea could help forecast bad weather. He believed that weather in one place could be predicted by judicious use of information from others.

Expanding the mandate of his office—to collect weather information from ships for maritime use—he'd set about establishing a system of spotters and telegraph messages to issue storm warnings for the nation.

By January 1990, human weather spotters were still being used, but the office also relied on maritime reports, satellite information and a basement supercomputer—a gigantic machine resembling a complex of gray file cabinets—which, when it was not helping meteorologists, was used by the national climatologists to model climate, in the same way Jim Hansen did in New York.

The meteorologists had watched for the past two days, as the innocuous-looking wave on a cold front in the north Atlantic, 400 miles southeast of Nova Scotia, had moved northeast at 50 knots an hour. Air pressure inside the already low-pressure area was plunging so fast—by at least one millibar an hour—that meteorologists labeled it a "bomb," a storm that can grow to hurricane strength in 12 to 24 hours, even from ordinary weather.

Meanwhile, rising winds had created a dangerous curling mass of clouds on satellite photos. The hooked shape was caused by the earth's rotation. If the earth didn't spin, wind would simply blow straight from high- to low-pressure areas and stop when the pressure equalized.

But since the earth is spinning, wind curves. It's called the "Coriolis effect."

Sure enough, soon after midnight on January 25, as the "bomb" approached the west coast of Ireland, NOAA-11 satellite shots showed a feathered pinwheel shape over the ocean, bearing down on the British Isles. The sharply delineated wheel meant the dangerous elements had combined.

Cornwall, on England's southwest coast, began experiencing strong gusts around 6:00 A.M. At Trengwainton Gardens, a National Trust forest, a third of the trees were blown over. By nightfall, 3 million more trees across Great Britain would be down.

The wind picked up a 120-foot-long chicken house in Cornwall and threw it 300 yards. Then it crushed 10,000 birds in three adjoining chicken houses.

Meteorologists would speculate that a tornado had hit the chicken houses. Fortunately, no humans had been around to see it.

So far, humans had been lucky.

That was about to change.

Swindon is a fast-growing English town roughly 80 miles west of London along the "M4 Corridor," called the silicon valley of England. Numerous computer firms have sprung up there, and Honda and Motorola factories attract even more skilled labor. Other residents commute to London.

It's a comfortable place, of shops and slate-roofed homes, none far from the old Roman Empire highway, paved now and still used, called Ermin Road. Julius Caesar himself led legionnaires up and down England on that road. At 12:40, Grange Junior School was open, about 300 yards from Ermin Road. Its students were ages 7 to 11.

Inside the two-story school, headmaster Malcolm Emery, a dark, bushy-eyebrowed, soft-spoken man of 48, was working in his ground floor office on the "French trip." On this day, when his life changed forever, Emery had been at his job for five years, and was planning on resigning, maybe going back to school to finish his masters degree, maybe enrolling in teacher training. He was popular with staff and students, a mild-mannered effective presence, but ready for a change.

And the "French trip" was an annual excursion to Normandy, for talented students. Emery decided to call the participating children into his office and work with them on their conversational French, which was not in the normal curriculum for students so young.

But looking outside at the bad weather, he changed his mind and decided to take a walk around the school and keep a "high profile." The 260 students had been kept inside all morning and had not been allowed onto the grassy field outside to play. His presence would calm them, he felt.

He began making rounds. The well-lit halls were decorated with student poems and paintings. Classroom doorways stood side by side, so Emery could see into two rooms at a time. Each had nice big windows, lots of space, and the children sat six at a table, boy beside girl, all in uniform: gray pants, white shirts and ties for the boys; gray skirts, kneesocks, white blouses and ties for the girls.

He could hear the wind outside. Occasionally it rattled windows in the 29-year-old building. It had been rising since he'd arrived this morning, from "a bit of a breeze," at 9:00 A.M., to the full storm it had become.

No rain yet. Just wind.

Satisfied that things were in hand, Emery went back to planning the French trip.

It was a high point of every school year.

Upstairs, in the southwest-corner classroom, trainee teacher Denise Randall conducted a nutrition lesson for a senior class. As she told them about healthy eating she looked out at a room where plants stood on windowsills and handwritten student autobiographies were affixed to walls. Metal weights and measures were on display, for math lessons. The class's usual teacher, Mary Read, had also exhibited student posters or dioramas on World War II, after teaching them about the London Blitz that year, explaining how children their own age had hidden beneath desks during air attacks.

The knowledge would come in handy a few minutes from now.

At the moment, Mary Read sat out in the corridor, scanning the new national curriculum she would soon be teaching, periodically glancing into the classroom to make sure things were calm. At one back table she saw ten-year-old Anthony More, a friendly, well-liked boy, whispering with Kiernan Smyth, known for his tendency to talk a lot, and with Paul Gallagher, son of a local restaurant owner. Emily McDonald, age 11, a clever, quiet girl, one of the tallest in class, also sat at the table, taking notes. She liked animals, and her wry sense of humor amused teachers.

Just recently, for instance, one of Emily's friends had approached Malcolm Emery in the schoolyard and told him, "Emily says she likes your tie."

"You tell Emily thank you," the flattered headmaster, not the nattiest dresser, said.

The child continued, sweetly. "Yes! Emily says it's a lot better than the tie you wore *yesterday*."

Now, as the children whispered or listened or took notes, the death and damage toll was rising across Great Britain. Wind gusts in places had reached 90 miles an hour.

In Cornwall, a hotel roof blew off, and injured guests were rushed to a hospital.

Near Falmouth, winds gusting to 100 miles per hour prevented two Royal Air Force Helicopters from leaving their hangars to help the 1,500-ton freighter *Celtic Navigator*, out at sea, listing dangerously because its timber cargo had shifted. Other helicopters had to be dispatched from farther away, and faced such strong winds that although they normally flew at 120 miles an hour, they could cover only 45 miles in their first 42 minutes of flight.

At Bristol, a 36-seat turboprop plane blew over on a runway, and ambulances were rushing to help free passengers trapped inside.

The M5 highway was closed along a 50-mile stretch between Bristol and Taunton, littered with high-sided trucks that had blown over.

Southeast of Swindon, in Hampshire County, 30 gliders that had been tethered to the ground broke free and were flying everywhere by themselves, out of control. Chief police inspector John Smith, father of five, was killed when a tree fell on his car as he drove to work.

All over the country, the storm's ferocity caught people by surprise. They knew what Malcolm Emery and Mary Read did. They lived in a place with lots of recorded history, and on land, at least, weather disasters weren't usually part of it.

People would say it later, all over Great Britain.

Things like this don't happen here.

———

Enjoying the unusual wind, Adrian Wys, 23, walked down a street near the school, eating fish and chips wrapped in waxed paper. On his day off, he'd driven into Swindon from his home in the nearby village of Christlade to get a haircut, but the shop was closed for lunch, and he was waiting for it to reopen.

Anyone looking at the baby-faced Wys would not have guessed he was a police constable about to become a local hero, not from his leather jacket and corduroy trousers. He was tall, strong and broad-shouldered, and his round face was topped with curly hair. He'd only been a cop for 15 months, and was just off job probation.

Still, a son and grandson of policemen, he had plenty of law enforcement experience. Wys was an ex-military cadet, and an ex-policeman in the Royal Air Force. He kept fit by running and canoeing, and he liked "messing about with cars and motorbikes."

As he walked, munching lunch, rubbish skidded down the road, and a milk carton blew past, five feet in the air. A responsible man, he chased it down and threw it in a trash can.

"Until that day I thought extreme weather would be fun. I was thinking, wow! The power of the wind! It never occurred to me that something horrible could happen as a result of wind."

After a while, though, the wind picked up, and grit began blowing in his eyes.

"It was getting past the fun stage."

He was less than a block from Grange Junior School, which was partially visible in gaps between homes, when, "from the corner of my left eye I saw a movement between two houses. I looked over. It seemed that I was looking at a roof settling into the top of a willow tree."

He thought he had seen the roof . . . well . . . floating.

But he knew such a thing could not *really* have happened.

"So I kept walking. It took a few moments for it to sink in."

A few steps later, the view was unobstructed.

"Shit," said Adrian Wys, starting to run toward the school.

The crazy thing was, even then, part of his mind thought the whole thing so unreal that he zipped the bag of fish and chips into his jacket, so he could finish it later. What he was seeing seemed so unbelievable that he felt stupid throwing away a perfectly good lunch.

———

Bill Collins, 40, watch commander for the Wiltshire County Fire Brigade, had spent the morning in a poorer area of Swindon, atop a public housing project, where firefighters had been trying to tie down a metal roof.

"Once or twice a year we get bad storms here, but this one was different. I grew up here and never saw a storm like this."

To make his job harder, the communication system was out, due to bad weather.

"In the past, if we had a phone problem, we used the radio. If we had a radio problem we could set up relay patch sets. This was the only time I recall losing everything."

Having finally tied down the troublesome roof, Collins was heading back to his firehouse, two minutes from the school. He was due to go off duty and planned to take his son to the dentist. He'd been a fireman for 24 years, since age 16, when he'd signed on to avoid taking the kind of railroad factory jobs all his friends were getting.

In fact he'd hitchhiked 25 miles just to take the interview.

"I didn't want to be a mechanic or a bricklayer. And I liked being a fireman from the day I joined."

Collins was still youthful, slim and fit with singer Paul McCartney looks. He'd been divorced for five years, and his two kids, a 12-year-old girl and a 10-year-old boy, affectionately called him an aging hippie. He was a vegetarian, a contributor to Greenpeace, and a future member of Amnesty International. A lover of nature, he enjoyed watching animal shows on television, but he was not an outdoorsman, and as a single parent didn't have much leisure time anyway. But he could spend hours at home reading any kind of books or listening to any kind of music: Rock and roll. Classical. Opera arias. While the kids did homework.

Obviously, over the years, at his job, he'd "seen a lot of unpleasant things. Dozens, maybe hundreds of people killed. The first job I ever went on involved a family being wiped out. Twin children and their parents. Their minivan had gone over an embankment. I don't want to sound callous but you see things and it doesn't mean a lot to you. You go into professional mode, probably like combat in a way. The biggest pileup I ever saw on the motorway was sixty-nine vehicles. Imagine turning up in your little fire engine and there's ten miles of vehicles. You have to go on automatic to deal with it.

"But we weren't trained for the kind of thing that happened at the Grange Junior School."

It would "confuse me, and devastate me."

Ten years later, Collins would still be shaking his head over it, saying, "I don't understand why."

Around the time Bill Collins reached the firehouse, to be told by waiting firefighters that there was a roof problem at Grange Junior School,
roughly 130 miles to the southeast, at the historic port of Dover, Captain James Martin was deciding whether to try to reach Calais, on the
coast of France, in the P&O ferry he commanded, the *Pride of Kent*.

He was scheduled to make three round trips today during his 12-
hour shift.

Watching from the top deck as passengers and cars advanced unsteadily
up ramps into the ship, between violent surges in the berth, Martin
decided that staying in Dover was out of the question. Even though a
breakwater shielded the harbor, "The ship was crashing up and down in
the berth, as they do in severe conditions. We were starting to do damage."

But that didn't mean he had to go all the way to Calais. He could, he
knew, order passengers and vehicles off the boat and proceed to a more
protected anchorage, five or six miles up the coast. "The Downs" had
protected British ships in storms for hundreds of years.

Go to Calais? Or the Downs?

The forecast, for force 10 winds, blowing roughly 56–63 miles per
hour, "was bad, but not enough to abort the voyage. There's an expres-
sion 'adverse weather,' which to a layman means conditions are truly
horrible. But to a seaman they're conditions which impinge upon the
voyage, inhibit your ability to get where you want to as quickly as you
would normally. But they're not conditions that would necessarily prevent you from arriving at your destination."

So Martin told his crew they'd head for Calais.

"Nothing suggested that a storm of historical magnitude was immi-
nent."

Besides, he was an experienced captain who'd worked for 40 years at
sea, 28 of them commanding channel ferries, and 5 of *them* on this very
ship. In all that time his roughest trip had occurred in 1959, in a hurricane, when, as a 17-year-old second-trip apprentice on the *Port Huon*,
off Bermuda, bound for Australia via the Panama Canal, he'd spent a
turgid three days in a tiny wheelhouse, admiring the way his captain,
lashed in a pilot chair, exhibited "great calm and an inexhaustible supply
of foul oaths."

And if Martin was a veteran, so was the *Pride of Kent*, a proud part of P&O's fleet, which safely ferried 4 million passengers in and out of Dover each year. The ship looked nothing like smaller harbor ferries common in the U.S. Built for sea work, the 21-year-old boat stretched 160 meters in length, weighed 20,000 tons and had eight decks, two of them just for cars and trucks. The three engines could produce 25,000 horsepower. It was a smart-looking carrier, its hull royal blue, its super-structure crisp white and blue, its funnels decorated with the 150-year-old P&O company logo: red, white, blue and gold triangles.

Now, as Martin completed last-minute preparations, his crew tied down truck axles with thick chains, to keep the lorries from breaking free and doing damage in rough seas.

They double-checked the heavy extra inner door designed to increase watertight integrity in the parking area.

On the upper decks, the passenger areas, tourists, businesspeople and shoppers bound for France made themselves as comfortable as possible in the restaurants, food court, bars, currency exchange and even in a duty-free shop rivaling ones found in international airports. Gambling machines glittered, ready to accommodate players seeking to win some extra cash during the expected 90-minute trip to Calais.

The lifeboats and open decks were checked. The galley and purser's staff were warned of severe conditions and a reassuring broadcast was made to the passengers.

But Captain Martin's first try at leaving the berth failed even with tug assistance. On the second attempt he got the ship clear. The wind blew at a steady 60 knots as he turned the *Pride of Kent* toward the harbor's eastern entrance, cleared the breakwater and reached the open sea at full speed, under full helm.

He ordered the ship's movable fin stabilizers extended, to cut down on rolling.

"Right about then, the forecast changed to force 11–12, which is really odd," he'd later say.

Force 12 winds, Martin knew, could reach anywhere upwards from 73 miles per hour to "off the scale."

He still had the option of taking the fully loaded ferry to the protected Downs. "But in these latitudes, to actually get those highest wind speeds over sustained periods is unusual . . . As the lower registers of force 10–12 (64–73 knots) were within the ship's proven operational capabilities, subject to wind direction and tidal conditions and tug assistance, it was considered feasible to head for Calais," he wrote in his log.

The ship was rolling heavily, and as he brought it around to a favorable heading, plates and cups began smashing in the restaurants. Shoppers in the duty-free shops watched as bottles crashed on deck.

Over the intercom, Captain Martin warned passengers of conditions and asked them to take care when moving about the ship.

He reduced speed and adjusted course to make the voyage as comfortable as possible.

He ordered crewmen to affix extra chains to the trucks in the vehicle area.

Coast Guard watchers stationed at Langdon Battery, atop the white cliffs of Dover, watched the *Pride of Kent* fade from view in the storm.

Mary Read, sixth grade teacher at Grange Junior School in Swindon, was a small, soft-spoken, bespectacled woman with an understated directness—the kind of person who, when furious, might say, quietly, "I am *very*, very cross." At 49, she loved her job and considered herself satisfied and grateful for "a privileged, comfortable life." Her affection for children was reciprocated by her students. After 15 years of teaching, she was familiar with the noises the school usually made in heavy wind.

"It creaked and banged."

But that afternoon, as she sat in the hallway outside her classroom, scanning the new national curriculum, "the wind was suddenly roaring more, a slow continuous roar that went up and up. And suddenly it was all coming down. White descending. I heard a noise and ran into the classroom. Things started to fall."

She saw children ducking under their desks.

"Later they told me they remembered the films about air raids during the war years."

Mary Read experienced a sense of space opening up above her, and suddenly the wind was *inside* the building, smashing her, and small objects: paper, pencils, sheets of handwritten student autobiographies, ripped from the walls, started to spin in an airborne vortex in the center of the room.

"I couldn't see. I couldn't breathe."

She and the student teacher managed to get the students out from under the desks.

"One boy was bleeding badly from his leg."

They began evacuating the children. In the hallway, Mary saw children streaming from other classrooms, heading for stairs on either end of the school. She got her group downstairs, although the bleeding boy, Paul Gallagher, had to stop in the stairway. Teachers clustered around him, trying to stop the flow of blood.

Meanwhile in the hall downstairs, "when we have a fire drill, we count the students and then I call out their names," she said. "So that's what I did. I sat them down. I started counting them. It seemed to take forever. I knew I had thirty-three students in that class but I could only get to thirty-one. And then somebody said one of the children had been taken to the medical room."

Which meant one was missing.

Mary Read heard a thin, high voice pipe up from one of the boys in her class.

"Mrs. Read," the boy called out, "Emily didn't move."

———————

For Malcolm Emery, headmaster, who'd been working on the French trip, the nightmare started with a crash that sounded like lunch trays falling in the building.

"Then I heard screaming."

He rushed into the first-floor hallway outside his office. He never forgot what he saw.

"In fire drills we insist on silence, but this was the real thing. The children were coming along perfectly, orderly, moving down the stairs, doing exactly what they were told. But they were screaming. Face after face, screaming. It went right through the school."

The juxtaposition of order and panic, the sight of the little bodies moving the way they had been trained to, as the little mouths screamed, accentuated the madness of what was happening.

"One of the teachers called to me. 'It's the roof!' "

The headmaster hurried into the stairway, where he found Mary

Read and another teacher bent over Paul Gallagher. Mary told Malcolm there was still a child in her classroom.

He kept going, and at the top of the landing, faced the room.

There was no ceiling inside, and as in the horror movie *Poltergeist*, objects were spinning in the middle of the room: books, papers, roofing bits, insulation, dust. Crashes reverberated as heavier construction material—beams or pieces of wall—impacted against the floor.

"I saw Emily right away."

She was still in her seat, at her desk. She looked like she was taking a nap. He could see part of her uniform: the gray skirt and white shirt. Her head was on the desk.

"She was curled around the table.

"I didn't dare move her. I thought she was dead. I touched her, gently. I couldn't find any sort of pulse. There was no point trying to lift her.

"It was like a scene from hell."

Then, to Malcolm's horror, he heard a voice from below calling, "There are *more* children missing!"

Nothing in his training had prepared him for this kind of crisis, and he was terrified.

"In the center of the classroom I saw some rubble. I thought, 'God, they're under there.' I didn't think there was anything I could do for Emily. I was sure as I could be that she was dead. And I thought, if I start moving her, I'm going to do more harm than good."

So he decided to try to find the missing children.

"I started pulling debris away. But I thought, if I see a child's face looking back at me, I don't know if I'm going to be able to carry through rescuing them. I thought, if I see a hand or a foot I've got some chance to psych myself to deal with it. I thought, 'I know it's irrational, but don't be a face. *Please. Don't let it be a face.*'"

Then Emery noticed another man standing in the room, and the man said he was a policeman.

The headmaster snapped, "For God's sake, don't stand in that spot! You could be squashing someone!"

But a voice was calling, somewhere else, "We found them!"

"I thought, at least thank God for that."

Malcolm was now going into shock.

———

Adrian Wys, overcoming his sense of unreality, had rushed to the school, found the gardener and made sure that the gas and electrical power were shut off.

"We'd had drills in the police on how to evacuate buildings, if there was a bombing or something like that. Weather was never mentioned."

He'd gone into the gardener's bungalow to try to phone police or the fire brigade.

"But the phone was out of order."

Wys had rushed into the street and stood, waving his arms, stopping cars, telling drivers to "get to a phone. Call the fire brigade. The roof has blown off the school."

Which is how the fire brigade had been alerted to what had occurred. By a driver.

Wys had also checked to make sure the teachers were taking roll call, and had noted that students were being evacuated in an orderly way to a community center several hundred yards away. He'd found Mary Read and Paul Gallagher on the stairs, and Malcolm Emery in the class-room.

"I've dreamt hundreds of times about what I saw in that room. It was odd, because we were in a building, but it was like being on the top of an open double-decker bus. I could see the girl in a fetal position. Her face looked normal. Her head looked normal. But her body wasn't normal. It was twisted. Both her legs were the wrong way around.

"I felt for a pulse underneath her jawline. There was nothing. I didn't trust myself to check the pulse on her wrist. I was afraid my own pulse might drown her out."

Adrian told Emery, as the headmaster watched, "I'm afraid she's passed away."

Emery whispered to Adrian, even though he'd already thought it, "She can't be . . ."

And later, after hearing what Bill Collins would say at the inquest, Adrian would torture himself, and wonder if Emery had been right.

Bill Collins was arriving at the school now, pulling into the driveway in the fire truck, with his four-man crew.

"That wind was the highest I ever remember."

As he climbed from the truck, he couldn't see any roof damage yet, but "the whole school was evacuating. Hundreds of kids."

It looked like chaos. Then he saw the children were being led into the community center.

But the same sense of unreality that had gripped Adrian Wys and Malcolm Emery overcame Collins.

"I told myself, 'Something probably fell off the roof and they're evacuating to be on the safe side.' "

Still, he sent one of his men to the school office, to try to find a working phone. He ordered a second firefighter to remain at the truck and try to transmit an assistance message. If the man couldn't get through, he was to find Collins, and help inside.

Malcolm Emery, having come downstairs, "told us about the child upstairs. That was when the lack of communication became more important. I could have called medics otherwise.

"I remember hundreds of children's faces going past."

In minutes Collins and his two remaining crew found Mary Read's second-floor classroom. The floor seemed to be tilting, he thought, and not only was a section of roof gone, but a back wall had partially crumbled. What remained of it, beyond blowing debris, was barely chest high, more of a jagged fence of masonry, and beyond Collins saw the incongruous sight of a field and trees.

"I thought, and this is going to sound daft, 'Isn't this odd? This isn't normal.' But of course it wasn't normal. That's why we were there."

Collins spotted Emily McDonald, off to the right.

"We went straight to her. *We always assume a person is alive.* We are not qualified to say they are dead. So we started taking stuff off her.

"Bearing in mind the probable nature of her injuries, we had to be careful about moving her. I thought, if we've got any chance of reviving her, it's got to be done now.

"The two men with me were both ex-army. Both served in Northern Ireland. They were worldly, not kids, and had seen an awful lot.

Between us, we were quite experienced, and they had both just requalified in first aid.

"Because of the way Emily was sitting we had to stretch her out—get her onto a flat surface—in order to do CPR. But the likelihood of causing more injury was too great if we tried to carry her to another part of the school. So I sat down in the rubble. I put my legs out. We slid her onto them. Her feet were by my feet. Her head was by my hips. The other men started the two-person revival procedure while I started a physical examination, feeling for broken ribs, broken bones.

"She wasn't breathing, and at one level I believe Emily died instantaneously. Probably she never knew what happened."

"But I got a pulse."

Ten years later, the scene would still torture and baffle Collins. Ten years later, he'd sit in his home near the school and have trouble talking about it.

"I know you can't get pulses in someone's back. I've been over this a million times. You don't get a pulse in the spine . . . but I *felt* one. I was shocked."

So Collins told the other men, "I have a pulse," and they redoubled their efforts.

For Collins, time faded. He had no idea if he was there for minutes, or longer. He was conscious only of the steady beat reverberating in his fingertips. Ba DUM. Ba DUM.

After a while, he looked up, and a man was there. The man said he was a doctor,

"I have a pulse," Collins said.

"No you don't," the doctor said.

The wind was blowing harder in the room. Gusts in Swindon would top 90 miles an hour that day.

Collins repeated, "Seriously, I *have a pulse*. There's nothing emotional about this. This is professional."

"It's your imagination," the doctor said.

So the firemen stopped doing CPR.

Collins would say, falteringly, years later, "I'm not disputing his decision . . . He's a doctor. I've never questioned a doctor. I don't know anyone who has. If a doctor says to you, *that person is dead,* you take their word for it. To question a doctor wouldn't have occurred to me.

"We covered her up. We waited for the ambulance to take her away."
Ten years later, the memory would still drive Collins crazy.
"I FELT A PULSE."

The storm, reaching London, brought it to a standstill. Winds smashed windows and advertising billboards and blew pedestrians off their feet. Police closed entrances to the London underground, and diverted traffic away from Waterloo Bridge, after a bus overturned on it.

At Waterloo station, one of the busiest rail stations in the capital, stunned travelers stood in disbelief before scrawled notices reading, "This station is closed until further notice," as glass broke inside the concourse and shards swept out at them.

London Ambulance service fielded over 560 calls, over ten times the normal rate, in three hours.

Police sealed streets close to scaffolding-clad buildings.

Firefighters had trouble maneuvering their trucks through blocked roads as they tried to reach homes where roofs, chimneys or scaffolding had collapsed.

Cross-channel ferry service was halted.

But James Martin and the *Pride of Kent* had still not arrived at Calais.

For Martin, the surprises had kept multiplying since his departure from Dover, when he'd arranged by radio for a sheltered berth in the French port. An hour and a half later, having made decent time rockily crossing the Channel, he had reached the buoy marking the entrance to Calais Roads, and Martin, assessing the steady 60-knot wind and fast tidal flow across the harbor entrance, had decided to wait until after high tide was over to reach his berth.

But by now the storm was starting to inflict damage on the continental side of the Channel. French officials radioed Martin that unfortunately, both of their harbor tugs were needed to assist the cargo ship *Quay Paul Devot*, whose moorings had broken in the harbor, in strong winds.

No tugs would be available to help the *Pride of Kent* for at least two more hours.

"So I decided to make more westing to get sea room in case of further deterioration in the weather," Martin wrote.

It turned out to be a wise decision. Shortly after he headed west again, the wind began gusting over 100 knots, "accompanied by severely restricted visibility with driving rain and wind-driven spray."

The ship began pitching wildly. White water broke over the bow.

The radar stopped working, probably, Martin thought, because the powerful wind kept the unit's scanner motor from operating.

Without radar, he was now blind.

"It was ferocious. It got up very, very quickly. The storm was passing through the Channel rather than farther north, as usually happens. We suffered rapidly changing sea conditions more normally associated with the Atlantic."

As the crew moved among the passengers to give them comfort, Martin notified the British Meteorological Office back in Bracknell of conditions.

"They're the worst weather conditions I have seen in well over twenty years of service," he said.

He reported the ship's position to the Dover Strait Coast Guard.

"They were surprised that any ferry was still out to sea."

The wind gusts rose to 105 knots from a brutal base speed of between 70 and 95 knots. The tide now turned against the wind, further increasing, in the shallow Channel, the size of the growing waves.

"Admiral Beaufort's description of sea surface effect in hurricane winds—'air filled with foam and spray and sea completely white with driving spray; visibility seriously affected, sea surface streaked with foam and wave crests being blown off by wind' complied very accurately with the conditions," the understated captain said.

It was now too late for the ship to change course and try to reach shelter along the English coast.

"We ran the risk of being set over on our beam ends if we tried."

And even if they *did* reach the Downs, the Coast Guard reported over the radio that in the allegedly safe harbor, space had become congested with ships, and some of them were dragging their anchors in the fierce wind.

Martin felt "relatively safe" sitting out the storm but many passengers were terrified.

By late afternoon, as the wind dropped to a steady 70 knots, Martin decided he could safely put about and head back toward France. After carefully watching the swell pattern and waiting for the biggest swell to pass, he turned the ship and advised his suffering passengers that as soon as conditions moderated, he would enter the harbor.

In the end, no passengers suffered injury, and cargo damage was confined to a single broken taillight on a car. But a ferry ride that was supposed to be as easy as a bus trip had turned into a nightmare, and disaster had been averted thanks to the skill of the captain and the quality of the ship.

Elsewhere that day, others were not so lucky. The storm killed 95 people across Britain, Ireland, France, Germany and Denmark. It wreaked $4.6 billion worth of insurance damage. And "the damage to the (overall) economy is many times greater than the insured damage," one insurance report read. "In economically developed regions, total damage is estimated at three to five times greater."

Daria, as the storm of January 25 was named, blew itself out, spinning, northeast, into Eastern Europe, but the basic conditions that had given birth to it had not abated in the Atlantic. Over the next four weeks, Europe barely started recovering from one storm as the next one hit. On January 27, as the British Army helped clear debris at Grange Junior School and Tory Party chairman Kenneth Baker visited the wreckage, and as politicians demanded investigations over the tragedy, the second big storm of the month swept across Britain, costing another $150 million in insured damage.

On February 4, three days after Emily McDonald's massively attended funeral, winter storm Herta swept across France, Luxembourg and Germany, killing 28, and raising the tally another $850 million in insured damage.

On February 7, as the 260 students from Grange Junior School prepared to begin lessons again, in a temporary school, miles away, Judith stormed out of the Atlantic, killing 10 more people and damaging $140 million worth of insured property in Great Britain, France, Germany and the Benelux countries.

The worst must be over, meteorologists thought.

It wasn't.

Nana struck on February 11, two days after engineers discovered that the Grange Junior School roof had collapsed because the connection of the roof trusses to the wall plates had been inadequate. Nana added another $80 million in insured damage to the destruction that had already been caused in Ireland, France and Great Britain.

Ottilie and Polly battered France, Benelux, Germany and Switzerland on February 13, around the time that Malcolm Emery noticed that at the new school, student play had become very violent since the first storm. The students were lining up and tackling each other in the schoolyard. Peaceful play had stopped.

"We didn't actually talk about the storm at first, while we tried to get the children to settle down in their new environment . . . tried to get them back into a school routine. We honestly weren't sure how to deal with the children, who were stressed, upset, unhappy. We didn't know how much to tell them."

Ottilie and Polly killed 30, raising Europe's insurance damage price another $350 million.

Vivian, sweeping out of the Atlantic on February 25, killed another 64, and ran up over $3.2 billion in insured damages across Ireland, Great Britain, Norway, Sweden, Finland, Denmark, France, Germany, Austria and Switzerland.

By now, at the new school, Malcolm Emery was finding that if a door slammed, students began screaming.

"But what made me decide that I had to address the situation was, one day students were washing paint trays upstairs, removing red paint, and there was a gully under the sinks and the paint backed up. One student said, 'That's blood.' I got a message in the office, that I better get upstairs. By the time I got there the rumor was, the blood was Emily's, and they'd seen her, in a nightdress, flying in the hall. They were hysterical. I got them together, and I said, 'This is enough! You silly children! You're letting your imaginations run away with you! You should be ashamed of yourselves!'

"I also started speaking to the children about the storm. When they got back to class, it was all they could talk about. There was a kind of collective sigh."

Surely the incredible rampage of storms *had* to be over for the year, Europeans thought.

They weren't over.

On February 28, as officials in Swindon wrestled with the dilemma of whether to rebuild or replace Grange Junior School, winter storm

Wiebke began its destructive rampage across Great Britain, France, Germany, Switzerland, Austria, the Netherlands and Italy, killing 15, and leaving $770 million in insured wreckage.

Among the interested parties tallying damage, aghast, was a group of specialists in Zurich, whose offices at 50/60 Mythenquai looked out at one of the most beautiful views any office worker could ever desire, a vision of nature at its best: Lake Zurich and the snowcapped Swiss Alps beyond.

But the view inside the office was less pleasant as the experts reviewed charts and graphs depicting the year's skyrocketing insurance losses. Swiss Re, their employer, is one of the world's largest "reinsurance" companies—corporations providing insurance to insurers more familiar to the public, like Aetna, State Farm, Geico, Travelers.

Which meant that Swiss Re and other reinsurance companies lost money any time catastrophic weather hit insured areas of the world.

Was it any wonder that Swiss Re had a whole department—the "economic studies department"—to monitor disasters anywhere on earth? Sitting in Zurich, in their white shirts and ties, or feminine business suits, department staff kept track of any time there was a train wreck, bridge collapse, landslide, earthquake, ferry sinking, airplane hijacking, rabies epidemic, fire in a mine, tornado, cold spell or any other kind of damage-producing incident that might cost Swiss Re money. The company needed to understand the nature and frequency of these disasters or it would be unable to determine profitable reinsurance rates.

Each year, from lovely Zurich, one of the safer places on earth when it came to cataclysmic weather, the economic studies department produced a report called "Sigma," detailing every insurance-loss-causing event on earth they could find.

In 1990, as the teachers, students and parents of Grange Junior School tried to make emotional sense of the tragedy that had befallen them, Swiss Re statisticians, poring over state documents, press reports, technical journals, insurance company reports and monetary compilation criteria, listed in purely numerical terms the toll from hundreds of disasters that had occurred that year.

In the Gulf of Mexico, when the supertanker *Mega Borg* caught fire off the coast of Texas, and spilled 11 million liters of crude oil into the sea, the report read, "Four dead. $16 million damage."

When a U.S. Air Force Lockheed C-5 Galaxy airplane crashed on takeoff, in Ramstein, Germany, because of motor damage, it showed up in the "Sigma" report too. "Thirteen dead. $50 million in damage."

The report noted a fire in a Naples plastics factory; a landslide in Croatia, an out-of-control truck that had driven into a funeral procession in India; a panic in a pedestrian tunnel in Mecca, Saudi Arabia, in which 1,426 people were trampled; a rash of deadly mushroom poisoning in Turkey; a death-producing flood in an Algerian zinc mine; a measles epidemic in Nigeria.

Compiling the information, the statisticians learned that aviation crashes had produced 5.6 percent of the insurance losses across earth that year. Major fires had killed 405 people and cost 8.5 percent of insured damage; bridge collapses, 0 percent; waterborne accidents, 2.7 percent.

But natural catastrophes had accounted for 82.5 percent of all insurance losses in 1990.

"1990 will go down in the history of insurance as the 'year of storms,' " editor J. Marbacher, head of economic studies, wrote in the report.

"Storm damage totalling $13 billion accounted for 76 percent of the total damage. Of this amount, $10.1 billion is attributable to a series of eight storms in Western Europe in the early part of the year, starting with Daria and ending with Wiebke."

The report also warned, "Catastrophes of this magnitude occurring in such a small area and over a short period of time put the system of insurance to a severe test."

The white shirts in Zurich were so concerned over that year's weather and what it meant that they commissioned a special report called "Storms over Europe."

Had the winter storms been a fluke, part of regularly occurring natural fluctuations in nature? Or did they represent some disturbing new trend?

For people whose livelihoods depended on calculating damage, this was more than an academic question. In 1990, for the second year running, insurance payouts from natural catastrophes had skyrocketed to levels well above what they had been through the 1970s and 1980s. And unlike those previous decades, when several years separated the big losses, insurance companies had paid out for "at least one such catastrophe per year since 1987."

Now, as Swiss Re tried to understand the factors affecting these losses, their report labeled Daria the quintessential "European Storm" and openly speculated about the role of the greenhouse effect on future storms.

Storms in Europe usually "arise along the polar front, the abrupt temperature borderline between cold polar and warm, humid subtropical air," the report said. The worst storms tend to occur from autumn into spring, when the polar area is colder and the moderate latitudes relatively warm.

"The greater the difference in temperature, the more formidable the depressions and also the storms."

The *path* of the storms, the report added, depends on deflecting pressure systems they encounter on the way. The result determines whether a "storm sweeps across Europe causing widespread damage or whether it is deflected northwards as it moves east."

What would happen to this whole system, the report speculated, "if the earth's atmosphere warms up?"

At this point, the uneasy Swiss Re writers said, they could envision two opposing results of global warming. In the first, polar temperatures would rise faster than temperatures in moderate latitudes, just as computer models like Jim Hansen's predicted. The temperature contrast along the polar front would drop, and storm intensity would happily decrease.

But in the other scenario, amid this mass of Rube Goldberg–like weather complication, "Storms which had traditionally moved across open sea to the north of the British Isles would now hit heavily populated areas to the south," because high-pressure systems would lose their strength to deflect the storms north anymore.

Which is what Daria had done when it hit Grange Junior School.

The white shirts in Zurich admitted that they were just speculating about the greenhouse effect. But for the first time, corporations were doing the speculating, not just scientists.

Suddenly insurance companies were growing *much* more interested in Tom Karl's work back in Asheville, and in the upcoming IPCC report.

The economic toll became part of the public record. The private emotional one of the weather disaster—always the more invisible cost—ate at the teachers, students and public servants who had been at Grange Junior School.

Malcolm Emery, headmaster, would not retire, but stay on and try to get the school up and running smoothly again.

"A captain doesn't desert his ship."

Emery "had to keep things together until we moved back to the rebuilt school, three years later."

When they finally moved back, Emery and the caretaker both started to think they heard children laughing when the school was deserted.

"I know it's a bit nutty. I heard it on the ground floor and the next floor as well."

Seven years later, suffering from headaches and other odd pains, Emery was asked by his doctor if he had experienced any stress recently. He couldn't think of any particular stressful situation but he did talk a bit about the storm at the school, years before.

"You have post-traumatic stress," the doctor said. "From that storm."

Three years after Daria tore the roof off her classroom, teacher Mary Read found herself at a family party. She thought she was over the effects of the storm.

"I didn't see my brother come up behind me and pop a champagne cork. Suddenly I screamed and screamed, and then I turned around and said," she remembered, softly, " 'You should tell me next time you're going to do that.'

"Emily was a beautiful girl, one of the tallest girls in class," Mary would say at the turn of the century. "I chose her to sit in the back because she was tall. And I will always feel that being tall, she couldn't get under the desk fast enough.

"Emily's mom used to walk her to school, to keep her safe. Afterwards, she said to me, 'If a school isn't safe, then the whole foundation of what you believe, of your security, isn't safe, is it?' "

Adrian Wys got back to police headquarters on January 25, 1990, found the bag of fish and chips still in his jacket, and threw it out. Before sending him home, his superiors insisted he see a psychological counselor.

"I had nothing to say to her."

The counselor asked Adrian what his "plan" was now. Adrian told her he was going to drink some beers and order Chinese food.

"Drinking is not the answer," the counselor said.

"I'm going to do it anyway," Adrian said.

But then he started having dreams about going into the room. And at the inquest, when he heard Bill Collins say that he'd found a pulse on the girl, "it upset me."

Adrian started wondering, was it possible that there *had* been a pulse? That he had missed it? That, arriving on the scene before Bill Collins, he'd been misled despite his training and knowledge? Was it possible that he could have revived Emily by doing mouth-to-mouth resuscitation?

As his worry grew, he received dozens of thank-you letters from students. He won a cash award from the police and used it to buy study materials for the sergeant's exam.

Adrian Wys, local hero, "felt guilty about profiting from a horrible experience, about not being upset at the time. It made me question my compassion."

He never talked about it until finally, one night, he found himself drinking beer with a friend who had served in the British Army, in the Falklands War, and in that pub the friend confessed that he'd always felt guilty *too*, because crazily, he'd felt he'd let down a friend who'd had his legs blown off in the Falklands . . . felt guilty that he hadn't been more sensitive to his friend.

Sharing experiences, Adrian started feeling better. He realized, the more he thought about it, that Emily McDonald had been dead when he reached her. There had never been any pulse. You can't find a pulse in someone's back.

"You can talk to a counselor all day, but if you tell someone who's undergone a shared experience, that goes a long way."

Bill Collins never stopped torturing himself over stopping CPR when the doctor told him Emily was dead. Years later, at a wedding, drinking with the other two firefighters who had been with him that day, "I detected there was something still there in them. I knew there was for me, but I didn't know how they felt. I thought, I'll show them a little bit of my vulnerability, so to speak. If they feel the same way I do, maybe they'll feel a little bit better to let go. So I said, 'Do you realize I had nightmares about this for months afterwards?' And they said, 'We did too.' I couldn't believe it. We'd all given ourselves a fight about this, without each of us knowing.

"It seems a bit daft, really."

But had the terrible storms of 1990, in any way at all—even in a broad statistical sense—been influenced by the greenhouse effect? Had greenhouse gasses even slightly increased the chances that Hurricane Daria and the other European storms that year had been born?

That was one practical question for Tom Karl, in North Carolina. He was finally ready to write his part of the 1990 IPCC report.

"Since the meeting at Dormy House we'd been working with the Russians, U.K., and people in the U.S. to get an assessment of the urban-heat-island issue."

Karl and the other scientists had decided, cross-checking their work, that the first of his big doubts about the greenhouse effect, the question of whether city "heat islands" were misleading scientists into thinking that the planet was warming—was for the most part solved.

The report being written would reflect his opinion. The earth was definitely warming.

But Karl's other concerns remained. He was still a skeptic over whether humans *caused* the warming, and over whether models like Jim Hansen's could predict future weather in any realistic way.

"We still didn't know if we had adequate models to get the sensitivity of the climate system correct."

As far as Karl was concerned, huge questions over the role of oceans in the climate system remained, as did problems relating to the accuracy of ocean temperature readings. And what about the role of other atmospheric gasses besides carbon dioxide? Did they enhance human contribution to the greenhouse effect, or mitigate it?

For instance, the role of sulfates, which *cool* the atmosphere, still weren't even included in calculations many computer models made.

"The IPCC report did not make the linkage between human activity and temperatures," Karl said.

To do that, in his mind, would have been irresponsible.

But the scientists did reach consensus that it was a "virtual certainty" that temperatures on the earth's surface would rise substantially in the next century, between 3 and 8 degrees Fahrenheit by 2050.

The next IPCC report, containing any further conclusions, wasn't due out for five more years.

Karl went back to work.

Meanwhile, meteorologists reported that 1990 was turning out to be one of the most violent periods of weather in the last 40 years. Just in the U.S., heat waves broke records in Phoenix and Los Angeles. The worst floods in decades devastated Texas, Oklahoma and Arkansas. Between the floods and 726 tornadoes to touch down across the Midwest that year, insurance damage reached billions of dollars.

Back in New York, Jim Hansen learned that he'd won his bet—that one year between 1990 and 1993 would turn out to be the warmest in earth's recorded history—on the first try.

Hugh Ellsaesser, the loser, sent a check and said that Hansen "just lucked out."

Hansen didn't cash the check.

Within months, nature would provide his computer model with a gigantic test of just how accurate it could be.

The Maldives ... the Philippines ... California

MAUMOON ABDUL GAYOOM, president of the Republic of the Maldives, arrived at his seafront office on the morning of May 30, 1991, to receive bad news.

"The Minister of Atolls has reported a very severe storm in the south," Gayoom's executive secretary said.

Not again, thought Gayoom.

"Get the minister," he said.

The storm was not particularly bad yet in the capital. From his second-floor window, the president saw light rain falling along Marine Drive, the main harbor street in Malé, the island seat of government in his archipelago nation. Gusts of light wind hit cars and scooters zipping along the cobblestones past five- and six-story government buildings, and they snapped the enormous national flag showing the Islamic crescent moon, bordered in red, set against a field of green, that dominated a small park near the water.

Inside a stone breakwater across the street, rain turned slick the gray hull of a Coast Guard launch anchored near Gayoom's yacht, a white cabin cruiser. Rain pelted docked "dhonis," Maldivian-designed low, wide fishing or cargo boats constructed from trunks of coconut trees.

"Mr. President, the minister has arrived."

Beyond the breakwater, water taxis crossed the rainswept atoll, bringing travelers to or from the nation's only international airport, on another island a mile away. Other covered boats ferried tourists between the capital and one of several resort islands further out.

Gayoom normally loved the view: the turquoise lagoon and green dots of islands beyond, the jets bringing vacationers from Europe and Asia, the big tankers anchoring or passing, as the Maldives sit on an ocean crossroads, on trade routes linking India, Africa, Asia and the Mideast.

He often stood at his window when he needed to think. But today the rain increased his apprehension over what might be going on over the horizon.

"Occasionally, during a storm, some trees will fall. Some houses will be damaged. It can happen during the southwest monsoon season. At first I thought it was that," President Gayoom said.

But now the Minister of Atoll Administration explained, frowning, that starting last night, and continuing into this morning, winds in parts of the country had reached over 100 miles an hour.

"No record existed of winds of that magnitude ever having hit the Maldives before."

Gayoom ordered his other ministers to assemble immediately for an emergency meeting. He instructed the Minister of Atoll Administration to stay in radio contact with the "atoll chiefs," appointed regional governors. But the command was more difficult than it sounded. The Republic of Maldives is more than 99 percent water. It consists of 1,200 islands scattered widely over 90,000 square kilometers of deep Indian Ocean, straddling the equator.

No island is larger than three square miles. Many are smaller than a New York City square block. The average island stands only five feet above sea level, making it enormously susceptible should oceans rise. The highest elevation in the country is only eight feet above wavelets lapping at the nearest shore—meaning that if worst-case IPCC scenarios were right, the whole country would disappear.

"We are so vulnerable. If any severe storm strikes us, it will do a lot of damage," Gayoom says.

In 1991 the country's 200 inhabited islands were reachable only by radio, that is, if they *had* a radio.

Should an emergency occur, the nearest country, India, lay 370 miles to the north.

On nice days—just about any other day—this visceral sense of isolation accentuated one of the most enticing views on earth. For Gayoom—or any air traveler—it was heightened whenever he flew over the
country, gazing down at atolls formed by necklaces of islands: elongated
ovals or curling spits encircling, in groups, the turquoise waters of their
shallow lagoons.

Every atoll was formed by a coral reef. Reefs and lagoons surrounded every island.

Newly forming islands appeared from the air as white sandy strips,
their submerged fringes bejeweled by the sea washing over them. Older
islands, fringed by beaches of fine white sand, were lush with tall
coconut palms, smaller breadfruit trees and screwpines.

No roads or bridges connect Maldives islands. No rivers cut through
them. No telephone poles and few roads intersect the islands, only three
of which include even small airports. Time can slow. Sea travelers can
take up to a restful month getting from one part of the country to
another by sailing dhoni.

What they see, when they reach inhabited islands, are villages constructed mostly of small homes built solidly from coral, with roofs of
beaten tin gleaming in the sun—although recently coral use in construction has been prohibited.

On special islands, government-demarcated tourist zones that Gayoom created, the view is more modern: of sophisticated vacation
resorts—where divers, sun worshipers and sport fishermen from around
the world enjoy the pleasures of the reefs, beach and sea, living in air-
conditioned huts, taking meals in fine restaurants, enjoying sundowners
(alcohol is not allowed elsewhere in the Muslim country) as they gape at
the kind of sunsets they'd once thought existed only on postcards.

The Maldives, in short, are the kind of peaceful place that would
have tempted Gauguin to jump ship, and more than a few ancestors of
its 250,000 citizens reached the country that way.

On May 30, 1991, as radio reports filtered in from atoll chiefs, Gayoom learned that the fishermen nearby, the tourist islands and the capital were at the moment out of danger, but other islands were not faring
so well. Homes were being destroyed, and valuable tree crops blown
down.

"I became more concerned."

He called the chief meteorologist of the country, his brother Majeed, while the cabinet hotly debated whether or not to alarm the citizens of Malé with forecasts for severe weather.

Gayoom also talked by radio to the atoll chiefs. The winds were too high for him to fly out and see damage at the moment, but he planned to do that when the storm died down.

"I thought what was happening might be a result of the processes of global warming," he said.

He'd read about global warming, had studied the theories. He'd already, in his mind, even had one life-threatening brush with it.

In fact, the only good part of the May 30 storm was that at least, so far, the unprecedented events of 1987 weren't repeating themselves.

At least, Gayoom thought, the terrifying high waves that had appeared out of nowhere in 1987 had not come back.

———

Maumoon Abdul Gayoom is a small, round-faced, bespectacled man who carries himself with a quiet formality and seems to consider each word before he speaks. Even on the equator, he dresses in western or tropical suits. His reticence is a subject of conversation in a country where people tend to be more easygoing, yet Gayoom seems popular if only judged from the fact that he usually brings only one or two police guards when he goes out. In person, he comes across as dignified, studious and slightly remote.

He loves cricket, astronomy, calligraphy and Agatha Christie novels. Son of a judge, Gayoom grew up when the Maldives was a sultanate and the only way to communicate outside the country was by telegraph. There were no banks. The law of the country was Sharia, the benevolently applied Islamic code. Gayoom, tenth of 25 children, got his first views of the other islands traveling by dhoni with his father.

"In those days the houses were huts with thatched roofs," said the president. "When we'd approach an island, the water would become shallower, and all of a sudden you'd start seeing fish. When you travel in deep sea you don't see anything, but the reefs are a spectacular sight. I would wait for it. It was a very affecting scene."

At the age of nine Gayoom won a scholarship to study abroad and was sent to Egypt, where after 15 years of intense work he was awarded a bachelor of arts degree in Islamic studies and law, graduating first in his class. He added a masters degree, spent two years lecturing in Islamic law and philosophy in Nigeria, and then returned home to teach.

By then the Maldives had become a republic, where the autocratic president, Ibrahim Nasir, ordered Gayoom arrested for criticizing the government in the privacy of his own home. Gayoom was banished by dhoni to a far island, where he lived with a fisherman for months, learning conditions and problems on the far islands. But in the flip-flop world of politics he was recalled, given a job by Nasir, and he rose to become UN Ambassador and then Minister of Transport.

When Nasir stepped down in 1978, Gayoom was elected president.

His record as a leader, by April 1987, when the first odd weather event occurred, was impressive, according to UN reports. Only one government school had been open in the atolls in 1978, when he took office. By 1987 there were 50. Life expectancy had risen from 47 years to 70. Malaria had been eradicated. Infant mortality had dropped from 120 per thousand to 30. Tourism was bringing millions in foreign exchange. And Gayoom was filled with plans to improve housing, health, food supply and clean-water supply.

By all measures, national conditions were improving when, on the evening of April 9, 1987, a new kind of worry appeared for the studious president: climate. Gayoom, weekending on his private island with his wife Nasreena, was sitting onshore, watching the sun set, when he saw in the usually clear water a first sign of what he now regards as the greatest problem his nation may ever face.

"I observed some murkiness."

He wondered what caused the murkiness, especially since the evening was calm. He and Nasreena returned to Malé by boat. He forgot about the oddly turgid water.

"The next morning, when I awoke, I was informed that part of Malé was underwater."

The crazy part was, the weather was still beautiful, cloudless and warm, but now waves had begun washing over the breakwater shielding the south side of the island.

"Nothing like this had happened before."

No old people remembered waves like these. No folk tales referred to them. In a country where the national calendar was divided into 24 "nakays," two-week-long seasons based on weather—a country where people regarded weather as part of their livelihood, culture and oral history, and so paid extra attention to its natural variations, no one could imagine where the waves came from.

At least they receded a little, so Gayoom relaxed, but that turned out to be premature. At noon a cabinet minister appeared anxiously in his office.

"They're back, higher than before."

Malé rises only five feet above sea level. On normal days, when waves are not topping the seawall, the low topography complements the postcard-perfect meshing of sea, land and climate, giving air travelers an almost miragelike view. The island—with its gold-roofed mosque; shore-road vista of fish market and National Security Service headquarters and government ministries; its neatly laid-out streets for Malé's 40,000 inhabitants—seems, all of it, somehow *lower* than the sea, as if some slender, invisible membrane keeps the water from rushing in.

The flip side, though, is that, were the sea to ever turn angry, anyone looking out *from* the capital, at the lone runway of the only international airport a mile away, would experience a severe sense of isolation, especially if, as Gayoom saw in 1987, the runway is topped by waves and the planes have flown away.

There's nothing like a submerged national airport—lights swept away, chunks of coral washing back and forth on the runway—to remind a president that if huge storms ever come to his island, if the "cyclone zone," which now lies a few hundred miles away from the Maldives, ever shifts, the shopkeepers and schoolchildren and businessmen and parents won't be able to pack their cars and drive across a causeway bridge to any mainland, as barrier-island refugees do in the U.S. They won't be able to fly away by going to an airport. They won't even be able to drive or walk into the interior of the island.

There isn't any interior. The whole island occupies one square mile.

"I was becoming worried," admitted President Gayoom.

So the president called the police chief, and together with the executive secretary, they drove to the southeast corner of the island, to see the damage closer-up.

"Rocks and stones were on the road. Waves were destroying the breakwater. Water was in houses. Rubble and garbage floated in the street."

The teahouses were underwater, and the soccer field.

"I was sitting in the front seat, beside the driver. Suddenly I heard a shout, 'There's a big wave coming!' "

Gayoom turned to see the wave hit the Jeep, and then he felt the vehicle being knocked sideways. Receding, the wave began to pull the Jeep and the president into the sea.

"It was an alarming situation."

Citizens who had been watching their president rushed forward and tried to restrain the Jeep. They fought off the wave and kept Gayoom safe.

"But two-thirds of Malé lay underwater."

Wet and shaken, Gayoom headed back to the office.

For three days, helplessly, he watched as, during high tides, the waves kept coming. Then, as mysteriously as they had come, the waves disappeared.

But would they come back? And would they be higher next time? *Would the cyclone zone shift?*

President Gayoom, by 1987, had read about the greenhouse effect. He'd been following the science and was familiar with predictions that sea level would rise dramatically by the middle of the twenty-first century. He understood also that if the sea *does* rise, it will not do so in neat, predictable increments, microscopically uniform spurts of barely noticed growth.

Scientific reports were clear on the subject: greenhouse-effect-caused sea rise would come with bigger storms, stronger winds.

Higher waves.

"The waves brought the matter home to me in a very harsh manner."

By 1991, when the unprecedented winds came, Gayoom had been worried about the waves for years. He had taken to calling his brother the meteorologist any time a possible severe storm approached, or a cyclone started up hundreds of miles away, in Asia.

"The president would ask if it could come here."

Someone from Gayoom's office would also phone Majeed if the daily weather report wasn't on Gayoom's desk by 11:30 each day. And meanwhile, Gayoom studied reports by USAID, by researchers at the University of Rhode Island and by a Japanese scientist, all trying to pinpoint *why*, on a perfectly calm day, freak waves had hit Malé.

The scientists had all discounted undersea earthquakes as a cause. The Japanese researcher had blamed the waves on a storm that had hit Australia, almost 5,000 kilometers away. The waves had traveled all that distance before striking Malé, he said. The Rhode Island report predicted that if high-end estimates of sea rise were accurate, "rising sea level will become an important or even dominant concern for the Maldives."

Which meant, Gayoom realized unhappily, that he was president of a country that had become a measuring stick of sea changes, a canary in the climate-change mine.

By 1991 Gayoom could simply *not stop thinking about those waves.* He'd told U.S. senator Tim Wirth about the waves at a meeting in Canada, and Wirth would still be telling the story, saying he'd never forget it, ten years later.

Gayoom had also sponsored an international meeting of small island states in the Maldives, and the participants were coordinating efforts to fight climate change.

Their organization, AOSIS, the Alliance of Small Island States, included 43 states and observers by 1991. Jamaica was a member. So was Fiji, Micronesia, Belize, Bahamas, Malta, Singapore, Tonga, Cuba and Suriname.

Individually, the tiny voices of these countries were sure to be lost in the great din of international politics. Together, they hoped they would be heard as they pushed at forums for the kind of treaty Margaret Thatcher had proposed the year before, an international agreement to limit carbon dioxide emissions across earth, which Gayoom hoped would be signed in Rio, at the Earth Summit, the following year.

But now, in Malé, weather had turned nasty again, and *again* Gayoom was getting bad news.

By afternoon that day, the reports of 100-mile-an-hour winds stopped, but that was because the anemometer blew off the collapsing weather station. The storm, passing, damaged over 3,400 homes and left 24,000 people, 10 percent of the country, homeless. Gayoom toured the damaged areas. He established a ministerial-level relief committee. The Ministry of Defense and National Security coordinated food and supply delivery. The Indian government sent air force transport planes and helicopters to begin delivering food.

In a country that lacked any formal disaster planning, a U.S. government analysis labeled Gayoom's handling of the crisis "remarkable" in speed and efficiency. It praised the way Gayoom's representatives selectively called foreign governments to ask for aid, instead of "blanketing the world" with major requests for assistance.

The Maldives republic was back to running smoothly in five months, although it would take years for tree crops to regenerate. And from now on, when Maldivian diplomats used the words "national defense," they would be talking about climate as well as military security.

When it came to climate, "we always have apprehension," said President Maumoon Abdul Gayoom.

———

Four weeks later, on June 10, 1991, fifteen thousand U.S. Air Force personnel, civilian employees and their families began streaming from homes and barracks at Clark Air Force Base on the Philippine island of Luzon, carrying babies, suitcases, food and pets. They crammed into waiting cars, trucks and buses.

As the volcano Mt. Pinatubo smoked 10 miles to the west, the convoys began evacuating the base, heading for American Subic Bay Naval Station, 50 miles away.

Mt. Pinatubo, dormant since 1360, had growled to life the day before, and was spewing rocks and hot ash down its steep slopes.

Already Clark's military aircraft had been moved to Cubi Point Naval Air Station. Now the convoys proceeded in the midday heat, in lines stretching up to two miles, through country occupied by Communist guerrillas. U.S. Marines and Philippine security forces patrolled the highways. U.S. helicopters patrolled overhead.

Volcanoes, of course, do not erupt because of the greenhouse effect. But Mt. Pinatubo was about to involve itself with Jim Hansen and in the global-warming debate in a big way.

On June 12 the mountain blew, shooting a mushroom cloud of gray-green ash and steam 50,000 feet into the air. Sand fell in the towns and villages nearby. Trees and streets were covered in ash the color of snow. Philippine Independence Day rallies become mass prayer sessions. Thousands of Aeta tribesmen fled on buffaloes and in ox carts as hot stones the size of cantaloupes fell at them.

Residents of Manila, 60 miles south, gaped at a rising plume 12 miles high.

Titanic eruptions continued into the next day, and by June 16, with satellites showing a 1.8-mile-long fissure cutting the top of the mountain, U.S. officials decided to evacuate Subic Bay base too. It was already covered with up to 12 inches of ash, and inch-long pebbles raining down.

Jim Hansen, in the U.S., learned more about the eruption, and the scientific opportunity it presented made him happy.

"It provided a great natural test of climate sensitivity. It would allow us to test our climate model."

Hansen knew that volcanoes spew gasses and dust into the air, and the gasses include sulfur dioxide, which forms sulfuric acid aerosols that winds can then spread around the planet into a blanket *cooling* the atmosphere, mitigating the greenhouse effect.

"I thought I could minimize confusion by publishing a prediction right away," said Hansen, who decided to use the climate model to project exactly how much the eruption would cool the atmosphere over the coming 18 months. If he could do that accurately, it would show the world that the model's way of processing information—not only about the volcano but about thousands of related aspects of climate—was correct.

"I'd make the cooling support the greenhouse theory."

He was taking a big risk, though. It was possible that other climatic influences or even a big swing in natural variation could cancel out any volcanic cooling effect. If Hansen's prediction failed, or was inaccurate, his whole argument would be in jeopardy.

Back in New York, "we created a special group. The Pinatubo group."

The group began collecting information necessary to make the prediction: crucial satellite measurements totaling the aerosols Mt. Pinatubo was discharging. In Hansen's computer model, the aerosols would ultimately become an outside influence, or "forcing," on normal climate dynamics—fundamental equations upon which the model operated and that represented the permanent dynamics of the atmosphere, modelers liked to claim.

Had the building been made of glass, shoppers looking up from busy

Broadway below would have seen the model, the whole operation in bits and pieces in rooms upstairs.

Hansen loved to tell the story of a high school student, visiting his office, who had once asked which room actually contained the climate model, so he could see it himself. Hansen had explained that the model wasn't like an airplane model, a construction made of wood or plastic, a tangible representation that you could hold in your hand. It was thousands of calculations.

The "model" was a grad student typing in one office, a PhD examining a satellite report in another, a scientist downloading ice core data from Greenland, or evaporation data gathered after thunderstorms, or weather record reports from Tom Karl's office in North Carolina.

The model then converted this data to numbers or equations representing basic climate dynamics.

To the model, the winds that blew across the plains of Montana, the snow that fell in the Andes, the shimmering heat that blasted camel caravans in Morocco were math equations representing the physics of energy absorption or expenditure in the atmosphere.

Long before Mt. Pinatubo erupted, the model had used as a starting point in its examination of "climate forcings," like the greenhouse effect, key basics of the atmosphere. Its ability to re-create them lay at the heart of the greenhouse dispute. The model "knew," for instance, that heat on earth started with the sun, where the temperature is 6,000 degrees Celsius, and that since the earth's equator tilts permanently *toward* the sun at the equator, equatorial regions are always warmer than the poles.

That was a given.

The model used as another starting point that air masses on earth move between warm and cold areas, seeking equilibrium, causing wind, which the spinning earth curves.

On and on. The model re-created, using numbers and physics, the way every single day of every year, warm air rises off the equator, and its departure creates a vacuum that cooler air from surrounding areas rushes in to fill. Sailors call this incoming air "trade winds," and know, as computer models do, that as trade winds move along the water they pick up moisture and that, rising, they shed the moisture as rain.

That's why it rains so much along the equator.

The computer model "knew" other basics too: like that eventually the rising, cooling equatorial air hits slightly warmer air in the stratosphere, and, blocked from rising further by that level, spreads out and falls back toward earth, warming and dryly reaching the planet along latitudes lying roughly 30 degrees north or south of the equator.

That's why most of the earth's major deserts lie in that region. The Sahara. The Mojave. The Kalahari.

Basic unchanging climate physics, the model presumed in its unemotional, robotic way, had driven earth's climate for thousands of years. The system's fundamentals were no different in 1991 than they had been when Moses led the Jews from Egypt, when Nero ruled Rome.

And by 1991, when Mt. Pinatubo erupted, computer modelers, not just in Jim Hansen's laboratory but at Princeton, at the Max Planck Institute in Germany, and in Boulder, Colorado, had been refining information fed into models. Models "knew" and incorporated into their projections that air in polar regions is dense and cold, and like air from a refrigerator, which oozes out as steam and spreads along the kitchen floor when a fourth grader opens the door, polar air oozes down the earth's surface, a gaseous glacier, veering east as the planet spins.

These winds are called "polar easterlies."

Models knew that boundaries between colder and warmer air masses, where bad weather tends to occur, are "jet streams," high wind caused by the moving collision zone between the two air masses.

"What the computer does, presumably, is solve fundamental equations that the atmosphere is obeying," explained David Rind, a scientist at Goddard. "These are conservation questions; of mass, momentum and energy.

"Say there's a fifty-degree temperature in New York City, on a particular day," Rind said. "The next day the temperature rises to sixty degrees. *Why did that happen?* There had to be an increase in the energy available over New York to warm the temperature. Just like when you turn on the stove and heat warms your pot. But in the case of New York, what could the energy be?

"Well, maybe yesterday was cloudy and today it's sunny. So more sunlight came in and warmed the temperature. Or maybe the air was blowing from the north yesterday, bringing colder temperatures, and today it's blowing from the south. The concept of 'conservation of energy' means that if temperature warms there has to be additional energy that was put into the system to do that.

"The same for momentum. We might have a wind in New York of, say, ten miles an hour, from the west one day. The next day it's twenty miles an hour. There had to be an increase in momentum for that to happen; velocity times mass to increase wind speed . . .

"The kind of physics I'm talking about—conservation of energy, mass and moisture—occurs everywhere, all over the globe, at all times, at all levels in the atmosphere. So if one year is warmer than the next, *there had to be extra energy put in the system that wasn't there before.*

"The model tries to solve the conservation equations numerically. We put basic equations in. We tell the model, hey, if there's more energy today, warm the temperature up. If there's less energy, cool it down. If warmer air comes from the south, that will warm temperature. If it's from the north and brings cooler air, that will cool temperature.

"The model tracks temperature, wind, rainfall. But in order to do that we have to describe to it, mathematically again, many physical processes going on in the atmosphere. For example, getting back to why it's warmer one day than the next, let's say there were less clouds on one of those days. The model would have to know how you can cause clouds. And one way clouds form is, you get lots of moisture converging in a region, and relative humidity up to one hundred percent. You get rainfall. You get clouds. So the model calculates how much rain is coming into the region and then it looks at a numerical code, and the code says, if moisture amount rises above a certain percent, *mathematically form a cloud* and have that cloud intercept sunlight, which is therefore no longer available at lower altitudes and therefore cools the temperature at lower levels."

If all this sounds complicated, that's one reason why diplomats and scientists argue about it. Modelers spend years trying to figure out how to represent, mathematically, the percentage of rainfall that infiltrates the ground versus the percentage that evaporates or runs into rivers. Modelers spend years studying the effects of heat or wind at different altitudes. Modelers are constantly trying to understand "feedbacks" in the system, the roles of oceans, water vapor or clouds when it comes to enhancing or diminishing the greenhouse effect.

When they think they understand the mix of these processes, and have finished programming the model to reflect them, the only way to test the model is to measure its predictions against real weather records.

In one kind of test, modelers program their computer to "predict" climate for periods that have already occurred *as if they have not occurred yet*, using information gathered from previous years. If the predictions match the real records, modelers grow more confident in the computer's ability to predict future climate too.

Jim Hansen now had a chance to test the model's thinking and predictive power in a different way. He would use it to predict short-term weather that had not happened yet. Unlike projections for the next century, which would require 50 or 100 more years of real weather observations to test, *this* test would have an answer in 18 months.

"Scientists don't jump up and down when they're excited, but I recognized that this was one of those opportunities that come along once in a decade."

The Pinatubo team learned from NASA satellites that Mt. Pinatubo had spewed 20 megatons of sulfur dioxide into the stratosphere.

"The forcing from the volcano was about four watts," Hansen concluded from the model. "That meant it reflected back about four watts of sunlight which otherwise would be absorbed by earth. An interesting number, because doubled CO_2 would cause a *warming* of four watts."

So Hansen published a prediction that the volcano aerosols would temporarily *more* than cancel out the effect of man-made greenhouse gasses. The earth would *cool* over the next year and a half by about half a degree Celsius. After that, the sulfur dioxide in the atmosphere would dissipate, and the earth would warm again.

He was betting again, only this time no money was on line.

All he risked was public disgrace. And treaty negotiations leading up to the Earth Summit could be crippled by further scientific doubt.

Peter Grey Scott and his wife, Teresa Ferguson, considered themselves happy, prosperous, comfortable, protected. They'd worked hard to build the smoothly running lifestyle they enjoyed. On the day of the fire, Sunday, October 20, 1991, as they hosted a picnic lunch for friends at Laguna Seca racetrack two and a half hours from their home in Oakland, California, they were surrounded by people, possessions and activities they loved, and were filled with a vast appreciative contentment.

Peter, 57, an architect, had left his firm and gone into private practice four years earlier, and now spent his many working hours designing California wineries, the Fairfield civic center and private homes. A small man, bespectacled, gentle and aesthetically minded, his hobbies ran to racing cars, a wine collection, fine art and photos. He also loved his superb stamp collection that he'd been putting together since the age of six.

To Peter, who had grown up in a family lacking money, where people worried about finances, possessions were more than objects. They were part of his aesthetic sensibility.

"I always wanted to experience the best that humans can do. I'm an architect and I care about everything very much. I want to select the silverware. The paintings. I believe we can survive and improve the world and that a person's contribution includes surrounding yourself with fine things."

Scott had started drawing things as a boy: motorcycles, houses, fine objects. "My taste always exceeded my ability to afford things. Drawing was a way to possess them. If you drew something realistically, you essentially had gotten the enjoyment out of it."

Teresa, his second wife, small, pretty and energetic, had told Peter, when they married, that she "wanted it all," and by 1991 felt she had found a way to balance the triple duty of career, motherhood and the shared responsibility of caring for Peter's invalid mother, who lived with them. Teresa worked as a project manager for an engineering firm, which in October 1991 was completing work on a container-shipping terminal for the port of Long Beach.

The tough job took her away from home for weeklong stretches. It was so hard that on that particular week she had burst into tears on arriving home on Friday night.

The family was completed by 14-year-old daughter Ginny, spending the weekend at a friend's home back in Oakland, 7-year-old Kate, who had come along to the racetrack, and by Peter's mother, Grandy, who, crippled by rheumatoid arthritis, remained confined to her bedroom on the lower level of their home on Alvarado Road, a lovely ridgetop drive on the border between Oakland and Berkeley, California. Peter had designed the home for his mother, moved into it after a divorce and purchased it when she was too ill to run it.

"Grandy was a very proper Scottish lady, imprisoned by her body," Teresa said. "If she walked at all, she was almost bent at a right angle. Even with her walker, she could barely get from her bed to her bathroom."

Aware of these severe limitations, Grandy worried sometimes, "what if I fell and no one knew I was here?" Or, "What if there were a fire? How would I get out?"

Fire particularly frightened her. Her grandson had died in a house fire.

But Peter and Teresa had arranged for constant care for Grandy. Caretakers stayed with her day and night, making meals for her, helping her move around, playing Scrabble with her and reading to her. In her painful declining years, at least Grandy had a loving environment. Kate spent hours with her grandmother, bringing pumpkin pie or cookies to her, or just talking with her as Grandy sat in the sun, in her wheelchair, and looked out from the back deck at the spectacular view—the thick wild brushland of Claremont Canyon, which separated the house from the University of California at Berkeley.

The beautiful rustic home, however, with its shed roof, 11-foot-high windows and hardwood floors, was situated in the kind of urban–wild interface area that Montana wildfire expert Don Despain worried about, the kind he thought might become much more susceptible to fires if climate change occurred. Many of the expensive homes nearby were constructed of attractive but flammable wood. They were surrounded by so much vegetation, especially combustible eucalyptus trees, that they were invisible from even the nearest neighbors. Eucalyptus trees emit highly flammable gasses when exposed to high temperatures. In fires, the gasses can cause the trees to explode.

Branches of large trees hung low over rooftops. Brush grew right beside homes along the narrow twisting streets. The wealthy residential areas of the East Bay Hills, on October 21, like the whole San Francisco Bay region, were in the fifth year of drought.

During the previous week, the normally cool, wet ocean breezes the area enjoys had been replaced by dry hot wind from the east. A large high-pressure mass had stalled over California's Central Valley, while a low-pressure zone sat offshore to the west. As any weather model "knew," wind had begun rushing from the high- to the low-pressure area in an attempt to equalize pressure. Anemometers on Mount Diablo, east of the city, recorded hot winds blowing toward Oakland at speeds over 40 miles an hour: winds similar to the Santa Ana winds that hit southern California.

On the previous afternoon, while Peter and Teresa happily watched car races in Laguna Seca, a small brush fire had begun less than a mile from their home, in a dry canyon between Buckingham Boulevard and Marlborough Terrace. Fans at California's Memorial Stadium in Berkeley, below, had looked up during the California Bears/University of Washington Huskies game to see dark smoke in the hills.

But Oakland firefighters had doused the flames and returned to their station that evening, leaving equipment at the scene so they could mop up the following morning. In the densely scrubby hills, covered with vegetation a yard deep in places, it was possible that the fire might have dug its way into the brush and could be waiting to re-erupt.

At 8:15 Sunday, they returned and began dealing with the hot zone until 10:53, when a gust of wind blew a spark into fresh dry vegetation just outside the burn area.

The fire flared. The winds worked like a bellows.

The third most costly fire in U.S. history had begun.

None of this was known by Peter and Teresa. At the Laguna Seca racetrack temperatures were hot but not alarmingly so. The weekend had been a social success. Saturday's picnic lunch had gone well, and the couple had spent last night at one of their favorite hotels, the Inn at Spanish Bay. They'd dressed up for dinner and relaxed with Kate at the restaurant, overlooking the Pacific. The previous week had been tough on Teresa, but she was unwinding, surrounded by friends and family. Peter had taken her hand across the table.

"What a difference a day makes," he'd said.

Sunday's lunch at the track had gone well too, and at five the Scotts loaded the food left over—deli meats, pasta salads, pop and iced tea— into the back of Teresa's BMW, and they started the two-and-a-half-hour drive home.

"Life seemed pretty good and I was full of contentment," Teresa said.

Then the car phone rang. Crystal, the nighttime house sitter—an unofficial sister in the family—was calling.

"Have you had your pager on?" Crystal asked Teresa as Peter drove.

"Obviously not."

"Have you been listening to the radio?"

"No. What's up?" Teresa said.

"Well, there's a huge fire in the Oakland Hills. The bad news is, we can't locate Grandy."

"I'm sure she's fine," Teresa said, calming Crystal, still basking in the easy mood of the weekend, assuming that the fire was contained, that Grandy's daytime caretaker was with her, that Crystal's worry was precautionary.

"Well, the police are telling us they evacuated everybody. But we checked and can't find her. We called the hospitals and we're about to check the shelters."

"Who's 'we'?" asked Teresa, growing more alarmed.

"My mother. Me. Some friends. Teresa, my mother just wants you to know, if you need a place to stay, you're welcome to stay with us."

"I'm sure that won't be necessary," Teresa said, thanked Crystal and instantly called Peter's sister Jane, who lived nearby, and was regarded as a competent "tough cookie" in the family.

Teresa was shocked to hear Jane crying hysterically. Teresa had never known her sister-in-law to cry.

"They can't find Mother," Jane wept. She told Teresa that every emergency operator she'd called had insisted that her mother had been evacuated. But Grandy wasn't with her caretaker. The daytime caretaker had arrived at the house that morning, made breakfast for Grandy and left briefly to attend church. While the caretaker was at services, police had blocked roads to the hills. They'd not allowed the caretaker to return to the house.

Jane told Teresa that the last time she had spoken to Grandy by phone had been around one. That the old woman had told Jane it was getting dark outside from smoke. That Grandy had said the 911 operator had advised her to get into the shower, turn on the water and wait to be rescued.

"Thanks for helping me," were Grandy's last words before she hung up.

"Grandy's probably been evacuated if the police say she has," Teresa told Jane, trying to control herself, too horrified to believe the worst. "And the fire's under control, isn't it?"

"No."

"Our *house* is okay, isn't it?"

"Teresa, your house is gone."

Now the real panic set in. All Teresa could think was, *I have to call Ginny. If there was a fire she'd try to get to her grandmother.*

But Ginny was safe, miles from the house, at the home of a friend. She was watching the fire from the friend's roof when Teresa phoned. She told Teresa, "I called 911 about Grandy but the operator wouldn't even take my address. They told me, 'it's under control, honey. Don't worry.' "

Peter, hearing only one side of the conversation, asked what was wrong. As he drove, his hands began to shake.

Kate began crying softly and Teresa silenced her, called Crystal back and repeated her earlier question. "I just want you to tell me one thing. Is there any chance at all my house is okay?"

"None."

But that was unbelievable. It could not be true. Sure, Crystal *said* it was true, but that didn't mean she was right. As Teresa began calling friends she asked *everyone* the same question, Was there was any chance the house was untouched?

Her boss told her, hopefully, "Well, they say on the radio that Alvarado Road is the demarcation line."

So maybe the house *was* okay, Teresa thought. And it was simply too awful to contemplate that something terrible had happened to Grandy.

For two tortuous hours, they continued north, listening to radio bulletins that grew progressively worse.

The fire, after flaring up, had gone out of control in the canyon formed by Claremont, Grizzly and Swainland ridges and Upper Rockridge hill. It had blown up toward Grizzly Peak Boulevard, and down toward Hiller Highlands at the same time. It had sent airborne flaming material across Highway 24, an eight-lane freeway, had burned almost 800 homes in the first hour, torching houses at one point at the rate of one every 11 seconds.

Only 20 minutes after the conflagration broke out, news reports said, the energy system had collapsed in the area because the wooden telephone poles caught fire as easily as the trees. The electric pumps filling the reservoirs that supplied water to firefighters had failed.

Sixteen people were dead so far. Some had been trapped in cars as they tried to escape down roads blocked by abandoned vehicles. Some had been burned alive trying to run. Some had been overcome by smoke.

Other fire departments had sent personnel to help.

After a while, Peter, Teresa and Kate could smell fire burning. As night fell they saw a glow ahead in the hills.

The sights and smells exacerbated their panic. As they reached Oakland they began stopping at hospitals and medical centers, asking if Peter's mother was there.

Grandy wasn't there.

They asked directions to the morgue. They had no idea where it was. They found themselves asking police they saw, "Where can we find the morgue?"

But Grandy wasn't there either, attendants said, adding that news report counts had been low when they said only 16 people had been killed by the fire.

"Oh, we have lots more than sixteen," one attendant said.

Peter and Teresa dropped off the extra picnic food at a Red Cross shelter and tried to reach their house, but they were stopped by police barricades below the hills.

"Fires were burning all over the place up there."

Cars and homeowners were massed at the barricades, trying to get up to their property. Peter left the car, skirted the roadblock, and, on foot, slipped up the dark twisty streets, which had been so desirable to homeowners only 24 hours before.

Police intercepted him and sent him back to the barricade.

Determined to get a view of the house, Peter snuck into brushy area near the barricade and began maneuvering through wildland and up streets, trying to work his way up.

Was it possible Grandy was still up there?

"I got stopped three times by homeowners with guns, demanding, '*What are you doing here?*'"

His own neighbors didn't know each other. They had never bothered to meet. Until today, many of them couldn't even see each other's homes, the vegetation was so thick.

Fortunately for Peter, one man recognized him, and the neighbors allowed him to pass. But when he finally got up the hill, he ran into a police car again.

"They stuffed me in the backseat and took me down."

Where was Grandy? The old Scottish lady, without her medication, could be dazed, or lost, and she would be terrified and disoriented.

That night, lying in a guest bedroom at a cousin's home, Peter told Teresa, grimly mocking his contentment of the night before, "What a difference a day makes."

The next morning, when Teresa felt herself waking, she kept her eyes closed, "and I thought, please God, let it not be true. Then I opened my eyes, and saw the light fixture above us. It was real."

Shortly after dawn, the family scattered, to try to find Peter's mother, and to try to reach the house. At Teresa's office, her boss provided someone to help full time in the search for her mother-in-law.

While the office functioned as search headquarters, a place where anyone could leave messages, Teresa tried to get to the house. Reaching the barricade, she asked a Berkeley policeman for assistance. When the policeman asked her address and replied, "I can't help you. You live in Oakland," she screamed, "Is that what you said yesterday when people wanted you to rescue my mother-in-law?"

There was a tap at her shoulder.

"I'll take you up," a cop standing behind her said.

And now finally, Teresa, heart thumping, was in a police car, cruising slowly up the twisting streets toward the ridge where her house might still be standing. The homes she passed were unburned at first.

"This isn't so bad," she said.

"Don't get your hopes up," the cop said.

They rounded a bend.

"Oh my God," she said.

Alvarado Road was a moonscape. "There were no reference points." There were no homes, or trees, or even *walls*. Cars were welded together. Houses had disintegrated. Lone freestanding chimneys rose above black wreckage, giving the scene the appearance of old Civil War photos: like Atlanta, burned by General Sherman, in his march to the sea.

Teresa saw that the only people on the street were sifting through ashes on her property, and the cop asked the people if they were "on a mission?" When they said yes, he advised Teresa that Grandy's dental records might be required for identification.

But the search squad didn't find Grandy.

Where *was* she?

Peter and Teresa tried the hospitals again, but by now the operators recognized their voices when they called and insisted, angrily, "We told you before! We have nobody by that name!"

When night fell they still had not found Grandy. News reports were announcing that 25 people had died, that 3,400 homes had burned.

The next day Peter and Teresa took a snapshot of Grandy to a news van in the hope of getting the photo on TV. Perhaps a viewer might recognize her. Amazingly, one of the newsmen eyed the likeness and said, "We just saw this person at a convalescence home. We have her on video! Wanna see?"

Peter and Teresa rushed into the van, where technicians ran the video.

It's *her*, Peter thought happily.

But at second glance, it wasn't her.

By now, the thrust of all official searches was for bodies. The fire had reached 2,000 degrees, reports said. One thousand, six hundred acres had burned. Gail Baxter, 61, had been found on the 6800 block of Charing Cross Road, killed trying to flee in the first minutes of the fire. Policeman John Grubensky, 32, had been trapped in his patrol car on Charing Cross Road, not far from Paul Tyrell, 61, found in his pickup truck. James Riley, 49, had been found near Kimberly Robson. Robert Emery Cox, 64, and Terry DuPont, 58, had been found in a backyard on Chancellor Place.

Governor Pete Wilson arrived to tour the damaged areas, and while he visited, roads were closed to homeowners again. But after he left, Peter, Teresa, Ginny and Kate finally all got up to the property—escorted by a homicide detective.

The police had not found Grandy. The rescue workers had not found Grandy. The newspeople had not found Grandy.

So the family began sifting in the ashes, looking for Grandy.

As they worked, finding lumps of ex-appliances, geodes that had cracked open, gravel that had once been on the roof—as they searched for anything recognizable, the TV van drove up with a Berkeley policeman, a big strawberry-blond Irishman who walked up to Teresa and said, "I'm really glad I found you." But the cop didn't look glad. He looked closer to tears.

The cop was "pretty sure," he said, that he was the last person to have visited the house before it burned. He said he'd been dispatched to rescue Grandy. He said, controlling his emotion, that when he'd reached the house, he'd looked in the back window and seen Grandy's unmade bed and her walker, and since she was nowhere in sight "he concluded that Oakland must have rescued her because they had been dispatched as well. It was somewhat of a logical conclusion," Teresa said.

The cop had banged on the door just in case Grandy was still in there. When he heard no answer, he'd left.

But fifteen minutes later, he'd gotten a *second* call to go to the house, that Grandy was still there.

"We broke down the front door. The walls were going in and out from the heat. The whole lower floor was on fire," the cop told Teresa, repeated that he was sorry, and left.

Peter said in a monotone, after that, "It doesn't get worse than this."

But it did get worse.

Teresa was standing at the opposite end of the house from Grandy's bedroom, and now she turned, and "right under my feet was a perfectly laid-out skeleton of, obviously, Grandy. I'd stepped into it. It was just an impression, the same color as ash. I picked up a bone and it disintegrated and at that moment it all came together. What we recognized was, Grandy didn't just die in the fire. She died struggling to get out.

"She had pulled herself the entire length of the house. She had reached a door but couldn't get out because it had a deadbolt. She couldn't turn it, and even if she had, where could she have gone?"

Grandy became the twenty-sixth name to appear on the roster of dead in the Oakland fire.

In the end, the fire was not officially brought under control until Wednesday morning. A hundred and fifty people were hospitalized, many suffering from serious burns or "oak poison" from a gas causing internal cauterization. Five thousand residents were homeless. Two thousand cars had been burned. Officials estimated insurance losses initially at over $1.2 billion, with overall economic losses expected to double that.

Private art collections like Peter Scott's, insured in excess of $18 million, were gone. City tax losses were estimated at $2.3 million, reforestation costs at $360 million. The vast losses to public roads and gas and water systems had not been calculated yet.

Damage from the Oakland fire was exceeded only by losses from the great Chicago Fire of 1871 and the San Francisco earthquake fire of 1906.

As the human and economic tally rose, so did accusations over why usual emergency procedures hadn't worked in this case. *Why* had the terrible fire happened? Why had the Oakland fire department cut personnel over the last several years? Why hadn't the department trained more in fighting urban–wild interface fires? Why had communication been so bad between fire departments, police departments, between emergency operators and the public? Why hadn't neighboring fire departments been called earlier? Why hadn't Grandy's caretaker been allowed to go back to the house?

As officials, government commissions and people like Peter Grey Scott sought answers, another quieter group of investigators arrived in Oakland and began driving through the blackened hills, talking to residents, officials, police and even weather forecasters.

The white shirts at Swiss Re, back in Zurich, had ordered the team in to figure out why the fire had occurred. For the *fourth year in a row* in 1991, the company was facing big insurance losses from a weather-related event.

"The fact is that a brush fire developed into a raging firestorm in a matter of minutes and proceeded to devour more than 3,000 houses in one of the world's most advanced nations," the report said, and cautioned that the Oakland fire "may be a harbinger of a new type of catastrophe that could reoccur on an even larger scale in the U.S., Japan, Western Europe or in other economically advanced industrial nations," primarily because of the way urban areas were being built in those places.

Why had the fire occurred? the report asked. "Like many other catastrophes, a complex chain and network of different influencing factors" had caused the massive loss. Three of the four prime factors were: the actual cause of the initial fire, the initially unsuccessful fire intervention and the suburban development structure.

But the fourth prime cause had been "climatic and weather conditions."

Not only that, but: "We cannot entirely rule out the possibility that the East Bay Hill fire was at least partially encouraged by global rises in temperature," the Swiss Re white shirts wrote.

Back in Zurich, that year, the economists were shaking their heads again over weather events that had caused insurance companies to hemorrhage losses.

Typhoon Mireille, hitting Japan, had cost a staggering $5.2 billion after it had oddly not lost strength when it made landfall, probably, Japanese insurers concluded, because the South China Sea, over which it had passed, had been warmer than usual.

Hurricane Bob, in the U.S., had cost $620 million.

Cyclone Gorky, hitting Bangladesh, had cost much less, $80 million in financial damage in a relatively uninsured area of the world. But it had left 10 million people homeless. It had left 140,000 people dead.

If the greenhouse effect leads to global warming, the Swiss Re report warned, "the atmosphere will contain more energy, causing a greater frequency and severity of storms."

Or had this process already started?

Of course, Swiss Re experts knew, Oakland, California, which was richer than any city in Bangladesh, or the Maldives, would always recover from weather disaster faster than places in the third world, would always be able to plan better for it, even in a post-greenhouse world.

But that did not mean that the developed world was immune to unpredictable disasters *or* that "fires of the future," as the report labeled the Oakland fire, wouldn't occur more frequently if greenhouse predictions were right.

Teresa Ferguson had a similar take on the subject.

"It is so easy to think that you can protect yourself by paying for security alarm systems, by having enough money to buy the right equipment. But none of that means anything in the long run. The Internet isn't going to help you if there's a major catastrophe. Neither will calling 911.

"Nobody wants to think they're vulnerable, especially to Mother Nature," she said.

———

In December, as thousands of Oakland residents tried to rebuild lives and grieve for loved ones, diplomats began arriving in Geneva, Switzerland, for the fourth round of negotiations designed to produce a working climate treaty to be signed at the Earth Summit in 1992.

Progress was not going well.

Delegates made their way through streets happily lit with Christmas lights and gathered in the Palais des Nations, a classic art-deco-style building overlooking Lake Geneva. Surrounded by UN security forces, the Palais lay inside barricades and barbed-wire fences, in a lovely compound of snow-blanketed lawn and conifer trees, all within view of Swiss villas and the nearby Intercontinental Hotel, where OPEC meetings take place and OPEC representatives like to stay.

The OPEC people opposed *any* treaty limiting oil or gas emissions. In negotiations, they stressed that disagreements among climate scientists cast doubt on whether any problem existed at all.

European representatives pushed for specific reduction targets, but U.S. representatives, dispatched by the Bush White House, reiterated long-standing opposition to them.

AOSIS countries like the Maldives thought proposed targets weren't high enough.

As the arguments grew bitter, even small points caused squabbles to erupt. For instance, at one meeting, after spending an hour and a half fighting over one paragraph in a proposed agreement, the chairman proposed creating a special diplomatic subgroup to deal with just that paragraph.

The Algerian delegate piped up, since there were 71 paragraphs, shouldn't there be 71 subgroups?

For those who believed there was urgency to the situation, the stall tactics were infuriating. Meanwhile, inside public meetings and in hallways outside private ones, gathered an increasing number of oil and coal company lobbyists. Energy and auto companies had formed the Washington-based "Global Climate Coalition," which was bullish on combatting any notion that their product could be linked to atmospheric degradation. Not only did they believe that the science was weak, but the legal climate in the U.S. worked against any urge an oil man might feel to change his mind.

As GCC spokesmen pointed out, when governments and companies had differences over policy in Europe, they worked them out in private. In the U.S., the government sued.

As the coal and oil lobbyists buttonholed diplomats in coffee shops and at cocktail parties, environmental organizations like Greenpeace sent their own representatives to argue for action. To them, the scientific uncertainties were minor and did not change general conclusions: that the earth was warming, that humans were responsible and that without a treaty, a cataclysm of global proportions loomed.

Even private conversations heightened the tension.

At one reception, for instance, over canapes, Greenpeace lobbyist Jeremy Leggett found himself talking with a Kuwaiti and an Iranian OPEC representative. Hoping he might influence them, Leggett pointed out that the British Meteorological Office had just confirmed that 1991 was the second hottest year on record. To him, this was more proof that the earth was in danger.

"Oh, our countries are accustomed to heat," the Kuwaiti replied.

Leggett then sarcastically asked the Iranian if he was "happy" that Iranian oil was adding so much carbon dioxide to the atmosphere. The Iranian smiled. "Can you *see* this carbon dioxide?" he asked, as if invisibility made it nonexistent, and he had scored a key conversational point.

As Leggett fumed inwardly the U.S. State Department representative joined the group, and Leggett asked him if the recent departure of John Sununu from the White House would make any difference in sluggish U.S. greenhouse policy.

"I think never," the State Department man said.

After a few more days of frustration, before Christmas, the fourth round of negotiations ended in disarray.

And as 1992—the year of the great Earth Summit—began, Jim Hansen was back in New York, staying out of politics, but working hard to refine his climate model.

The funny thing was, to prove global warming, he was now betting on the atmosphere to *cool*.

Jerusalem . . . Washington . . . Rio . . . Miami

IT WAS IMPOSSIBLE FOR ANYONE in Jerusalem not to notice. Snowfalls had always been an occasional possibility in the holy city, but the New Year's Day storm of 1992 showed no sign of ending. Shopkeepers closed early as drifts blanketed the bazaars of the old city. Muslim worshipers leaving the Dome of the Rock . . . Christian pilgrims following the route of Christ's last walk down the Via Dolorosa . . . and Jews wedging scribbled prayers into cracks in the Wailing Wall pulled their coats close and hurried back to homes or hotels.

The city ground to a halt. Falling trees knocked over electric lines. Drifts blocked the highway to Tel Aviv. Snow fell for the first reported time in the Negev Desert north of the Red Sea resort of Eilat.

But the assault kept coming throughout the astonished Mideast—coating olive groves, shepherds, trucks bringing oil from derricks to harbors.

Storms paralyzed Jordan's capital of Amman, and blanketed mountains in the Tobuk region of Saudi Arabia.

Blizzards shut down Beirut, Lebanon, for the first time in 40 years. They downed power lines in Damascus, capital of Syria. They isolated villages in Turkey, and set off avalanches in the mountains, killing 22.

The most brutal winter on record was beginning across North Africa and southwestern Asia.

In China, mercury plunged as low as minus 31 degrees Celsius.

In northwestern Iran, temperatures reached historic lows as that country's climate treaty negotiators got the news at the UN. The final rounds of negotiations were beginning before the Rio Summit, where whatever document was produced would be signed.

Global *warming*? smirked the OPEC diplomats, telling stories about the brutal snowfalls.

What a laugh.

––––––––––

The lights went out.

The screen lit up.

The video began with a pastoral scene, an oil painting of a shepherd and his flock in a lovely forest, during a pre-industrial time. The green leafy world on screen had no oil wells, electric power plants, or pollution-spewing cars.

Frederick Palmer, Washington lobbyist for Western Fuels Association, leaned forward, fascinated.

"Dawn of the industrial revolution," said a neutral-sounding female narrator, like an objective newscaster, which she was not. "Western Society is primarily agricultural and commercial. The concentration of carbon dioxide in the atmosphere is 270 parts per million. And then, factories spring up across the countryside and industrial towns grow around them."

The video, which Western Fuels had funded, showed another painting, this one of an old factory town.

"We begin to burn fossil fuels in large quantities—and then burn more," the narrator said.

Onscreen, an old newsreel showed sparks flying in a factory.

"The carbon dioxide level in the atmosphere rises."

Palmer watched a long line of early-model cars starting up. A prop plane took off, trailing smoke.

"The 1950s," the narrator said. "The concentration of carbon dioxide in the air is 315 parts per million. Still more fossil fuels are burned. More and more carbon dioxide is emitted into the atmosphere. More industrialization . . . More carbon dioxide . . . The year 2085. The atmospheric level of carbon dioxide has doubled to 540 parts per million."

"*What kind of world have we created?*" said the narrator, as the video showed an environmentalist's nightmare: hundreds of carbon-dioxide-spitting cars, on highways, in parking lots, in choking traffic, in gas stations, where they would be filled with more gasoline.

This is *good*, thought Palmer, who had read the script and knew what was coming.

The blunt-sounding video had certainly posed the question dramatically. Now Dr. Herman Mayeux, of the U.S. Department of Agriculture's Agricultural Research Service, appeared onscreen to give the surprise answer.

High levels of carbon dioxide will produce "a better world. A more productive world. Plants are the basis for all productivity on earth," he said pleasantly. "They're the only organisms that can utilize the sun's energy and create matter, food. And they're going to do that much more effectively, much more efficiently."

This is great, thought Palmer, as Western Fuels ratcheted up its fight against the climate treaty.

A second scientist—Dr. Bruce Kimball of the Water Conservation Laboratory, U.S. Department of Agriculture—appeared, with a green cotton field behind him.

"With a doubling of CO_2—cotton growers can look forward to yields that are sixty percent and more greater than what they are at present-day levels," he said.

Dr. Mary Brakke of the Botany Department, University of Florida, added, "For citrus [doubled CO_2] would be a very positive thing."

Dr. Gerd-Rainer Weber said, "Our world will be a much better one."

Finally the handsome Dr. Sherwood Idso, who had written a book on the subject, summed up the video's main point. "A doubling of CO_2 content of the atmosphere will produce a tremendous greening of planet earth."

The idea for "The Greening of Planet Earth" had been Palmer's, coming one evening when he'd watched a PBS special "After the Warming," at home, in his finished basement.

"It was one hundred percent apocalypse," he remembered in disgust.

To Palmer, the show had seemed to blame the coal and oil industries for an upcoming cataclysm. As he stomped around the basement, furious at what he regarded as irresponsible scaremongering, his eyes had fallen on the video *Cosmos*, a film version of Carl Sagan's book.

"I thought, Sagan popularized *his* book through video, and I liked Idso's book. It had affirmed the positive, optimistic view of people's impact on environment."

The more he considered it, the more he liked it. He contacted a filmmaker in New York and approved an outline for a film. Then he raised $250,000 each from the Western Fuels Association and the National Coal Association.

"We purchased the rights to Idso's book."

And now, in 1992, Palmer made the video available to journalists, politicians, corporate boards, Kiwanis clubs.

"Rush Limbaugh (the conservative broadcaster) picked up on the subject, and so did Paul Harvey. You could hear the phraseology from the video coming back over the air. And when *those* guys picked it up," Palmer said, delighted, "did *that* have an effect!"

The battle was heating up.

———

By May, at the UN, with only a month left until the Rio Summit, negotiators had still not produced a treaty. They knew that whatever happened in Brazil would be mere theater. The real decision to make the treaty meaningful would occur now, and the lobbying grew frantic as a result.

Jeremy Leggett, Greenpeace representative at the talks, argued that a historic opportunity was slipping away. But his opposites, like Washington lawyer Don Pearlman, feared that any document acknowledging global warming's existence could become a first step toward global fuel-use regulation.

Pearlman was a partner in the Washington, D.C., law and lobbying firm of Patton, Boggs and Blow, and founder of a nonprofit organization called the Climate Council, which represented power and coal companies. By securing official NGO (nongovernment organization) status for the council, he'd gained access to briefings from the official U.S. delegation. The slow-talking ex-chief-of-staff for the Secretaries of Energy and Interior for the Reagan administration had a jowly face, giving him a vague resemblance to the cartoon basset hound Deputy Dawg, and he

THE COMING STORM · 125

was seen frequently in hallways advising Saudi diplomats, poring over treaty language and buttonholing U.S. representatives to argue his point of view.

"How come there's not one line, not even one sentence on population?" he asked reporters, changing the subject when climate change came up, in his slow, jury-address-style voice. "I find it absolutely fascinating that of all the suggestions that have been made for helpful policies to deal with alleged global climate change, not one word on population . . . Never *once* do they put in a *sentence*," he'd continue, looking shocked, "or a paragraph or serious proposal for an international treaty which even *alludes* to the population issue."

Another lobbying presence at the UN was John Schlaes, new executive director of the Global Climate Coalition, whose members included Mobil, BP, Shell and Ford. The GCC distributed briefing sheets to delegates, debunking climate science, and held a press conference where University of Virginia scientist Fred Singer took issue with warnings about the greenhouse effect.

Positions were hardening. For every skeptic's argument about "feedbacks" that might mitigate global warming, like cooling sulfate gasses from factories, believers argued that feedbacks could work the other way too, *increasing* warming. Take methane hydrates, they'd say, referring to the crystalized high-pressure mix of water and methane found, in cold climates, beneath tundra, sediment and arctic seawater.

If temperatures rise enough to thaw out tundra, the methane hydrates will be released, some scientists feared. And with an estimated 10,000 billion tons of carbon stored in methane hydrate reserves around the planet, according to the U.S. geological survey, the extra global warming their release would cause might set in motion a runaway greenhouse effect, making current worst-case scenarios look mild.

The possibilities were so dangerous, European negotiators argued, that wasn't it logical to buy insurance against it? *That insurance would be the treaty!* Didn't the IPCC report, the record-breaking weather and the computer model predictions, taken together, *prove* there was reason to take out insurance?

No, the heated answer came back. It's all a coordinated effort to cripple the energy industry and impose international control on corporations and governments.

As the delegates fought, any one of them could stroll onto Second Avenue, into a book store, and look for a copy of a new book by Senator Al Gore of Tennessee. *Earth in the Balance* was a passionate treatise on the environment and global warming. Gore had held hearings for years on climate, and had warned about the greenhouse effect so much during his run for the Democratic Party presidential nomination in 1988 that one opponent had retorted that he sounded like he was running for national scientist.

In *Earth in the Balance*, he'd called global warming "the most serious threat we have ever faced."

He'd written, "In Tennessee, there is an old saying. When you are in a hole, stop digging . . . the truly conservative approach to the problem of global warming, for example, would be to stop thickening the blanket of greenhouse gasses and try to prevent further damage while we study our options."

Gore's book summed up the way the future vice president viewed people like Don Pearlman, Fred Palmer and OPEC negotiators.

"We now face the prospect of a kind of global civil war between those who refuse to consider the consequences of civilization's relentless advance and those who refuse to be silent partners in its destruction. The time has come to make this struggle the central organizing principle of world civilization."

In short, everyone battling on both sides had the same understanding of larger political consequences of the fight. If real, global warming posed a danger that nations had never experienced. Unlike nuclear threats, which came from a handful of countries, greenhouse gasses came from *all* countries, and from daily, average activities all over the world. Carbon dioxide didn't come from top-secret military laboratories but from millions of people driving cars, or turning on their stereo on Sunday morning. It came from buying air conditioners and taking vacations. It came from buying a new pair of shoes, made in an energy-consuming factory. Just turning on a lamp to read before going to bed added to global warming. Providing residents of a Pakistani village with basic electricity added to global warming, as usually a coal-fired power plant produced the light.

Which meant, if humans were ever going to curb their contribution to global warming, international law might eventually have to become more powerful to deal with it.

And *that* meant that global warming could become an issue that fundamentally altered power reserved for governments since the dawn of nation states—the power to regulate their own environment.

But even if international law *didn't* change, Al Gore's opponents feared that if ever given the chance, he would use the climate issue to try to institute energy taxes, more government controls over industries and restrictive laws limiting the lifestyle of average citizens.

Meanwhile, Gore's views about the environment would capture the public's attention in 1992 and the interest of Democratic Party presidential hopeful Bill Clinton, who would ask Gore to join his ticket as candidate for vice president; a big reason being, Clinton's senior adviser Rahm Emanuel said, "I always believed it was Gore's writing and thoughtfulness that caught his attention."

Gore would respond, before deciding whether to accept the offer, "I want to talk to you about the environment," and would show Clinton charts on global warming.

Then Bill Clinton would ask Gore *again* to be his running mate, and this time Gore would say yes.

At the White House, in the months leading up to the Rio Conference, as embattled President George Bush decided whether even to attend, Bill Reilly, Bush's head of EPA, was growing depressed. The president's head environmentalist found himself fighting over climate in meeting after meeting with John Sununu, powerful White House chief of staff.

"Climate was like nothing else in the Bush administration," said Reilly, a pleasant, mild-mannered, plain-spoken man. "It was a subject that raised enormous passion, anxiety and attention. By God, we had meetings coming out of our ears at the highest level!"

At the meetings, Bush or other top administration officials listened while Reilly and Sununu fought.

"I never did like John Sununu," Reilly said.

The men did not get along. One time, Reilly said, he went behind Sununu's back on an ozone limitation treaty question, phoned Margaret

Thatcher, and Thatcher convinced Bush to change his mind. The *Washington Post* broke the story, and Sununu, enraged, called Reilly at his farm in Virginia.

"Sununu said, your press office is out of control. I'm sending people over to take charge of it. I told Sununu, if you send 'em over, I'll be gone."

Sununu didn't send them over, Reilly said.

In front of the president, who preferred cordial disagreement to overt warfare, "We had a peculiar relationship in that we were wary of each other," Reilly said. "Sununu knew the president liked me. I knew he was chief of staff and had the president's ear. It was like we were in this little minuet which had us playing a certain role and not going outside the prescribed formalities."

Reilly, of course, wanted the U.S. to commit to specific cuts in carbon dioxide use, and specific dates by which the cuts must occur. He'd show up at meetings with EPA studies showing it was possible to tax energy and still enjoy "a net benefit to the economy."

Sununu didn't buy it. "Reilly was predicting doomsday with a computer model that neglected one of the most significant components to be modeled," he said, meaning the role of oceans in climate. "I kept saying, how can you guys make policy based on a model that is dead wrong?"

Reilly would quote from the IPCC report. He'd report to Bush regularly after traveling overseas and meeting with world leaders who supported targets and timetables. Chancellor Helmut Kohl of Germany was one of them.

"Kohl said to me, 'What I don't understand is, how Germany, which emits far less per capita greenhouse gasses than Americans do, can get reductions and you cannot.' "

Reilly also continually reminded Bush of his own promise made during the 1988 presidential campaign, to counter the greenhouse effect with a "White House effect."

Sununu felt that Reilly was too loyal to the small-is-beautiful crowd. "I thought Reilly came with too much of a commitment to a constituency that wanted to limit growth . . . I kept reminding him that it was the president who determined where the policy should end up. Not Bill Reilly."

"Sununu," Reilly said, "in meetings, could be very impressive. You engaged him with care."

Also, lots of other people in the administration felt as Sununu did. One time, at a meeting, a State Department official said to Bush, of the climate treaty, "Remember, Mr. President, this is a bet-your-economy decision."

"I thought, oh, great," Reilly said.

Another time, before a meeting, Reilly consulted with a State Department go-between with European countries on the treaty.

The man "laughed and told me, 'frankly, over here, climate change is a career killer. It's all yours. When it comes up tomorrow, I'll defer to you.'"

Away from the president, Reilly's frustration broke out as little jokes at Sununu's expense. Like the time, he said, they were at a gas station in a car, at a presidential photo op. Bush was trying to pump methanol fuel into a car. The visit was designed to promote the Clean Air Act, of which Reilly was enormously proud.

But Bush was unable to make the gasoline pump work, and there was a chance the photo op would backfire.

"I told Sununu, let's get that photo of the president off the evening news. You step over to the other pump and I'll take the nozzle off, and pour it on your shoe. *We'll* be on television, not him."

Reilly joked, "It'll establish to everyone's satisfaction that methanol is safe."

Sununu didn't go to the other pump.

"Sununu had a sense of humor, just not about himself," Reilly said.

"I'm surprised he admits to these childish acts," Sununu said, adding that he did not even remember the joke.

But jokes didn't help convince the president to support targets in a climate change treaty, or any biodiversity treaty at all being negotiated for Rio. In fact, six months before the summit, Bush had dispatched Reilly to Brazil, where he'd boarded the yacht *Britannia* for a cruise up the Amazon with President Fernando Collor de Mello of Brazil and Prince Charles of Britain. Reilly's job was to lay out the terms under which Bush would come to Rio, during the Earth Summit.

"Bush did not want to be embarrassed," Reilly said.

"Maybe sending him was a good way to get Reilly out of town," said Sununu, who back home was trying to keep Bush from going to

Rio at all. "There would have been no conference of magnitude without the president there."

As the *Britannia* cruised past the jungle, Reilly, Collor and Prince Charles discussed what specific wording in the climate and biodiversity treaties might be acceptable to George Bush. Reilly pushed for ambiguous language.

The Brazilian Minister of Energy told Reilly, "If you have a treaty which the U.S. hasn't signed, you don't really have a treaty."

Reilly agreed. Returning to Washington, he got to work trying to come up with language for the treaties. As spring of 1992 began, he still had not given up trying to convince Bush that targets and timetables were a good idea. And with Sununu gone by now from the White House, Reilly hoped his rival's departure would provide an opening.

"But I lost. When I walked out of that meeting where I knew everything was over on climate change, I reflected that from the president's perspective, he had his chief of staff, his vice president, his budget director, his chairman of the council of economic advisers, the chemical industry, the oil industry, the auto industry, the aluminum industry, the transportation industry, *all* on one side of the issue, and his EPA administrator and the environmentalists on the other. He had a recession. He had Pat Buchanan driving him crazy on the right, saying we were re-regulating the economy, and the *Wall Street Journal* saying the same.

"I thought, I shouldn't be surprised."

A reporter asked Reilly why, with so much weight on one side of the issue, Bush had even considered the other side.

"The rest of the world," Reilly said.

————

The document finally hammered out in New York for signing in Rio included language that "created a moral obligation but literally did not obligate us to targets and timetables," Reilly said. "There were no sanctions. It didn't require you to do anything. It was considered a victory by the industrial leadership of the U.S."

The negotiators had drawn up a Climate "Convention." Too weak to constitute a real treaty, it was more a statement of principles about global warming and a framework for actions to be filled in later, if the squabbling interests could ever agree on what they might include.

Grandly written, the convention set an "ultimate objective" of stabilizing greenhouse gas concentrations in the atmosphere at a level that would prevent dangerous interference with the climate system.

It did not specify what those concentrations could be.

The convention directed that "such a level should be achieved within a time frame sufficient to allow ecosystems to adapt naturally to climate change, to ensure that food production is not threatened and to enable economic development to proceed in a sustainable manner."

It did not specify what such a time frame might be.

It directed that countries ratifying the document take climate change into account when formulating policies on energy, natural resources and sea coasts.

It did not specify *how* to do this.

The convention also directed that signatory countries develop programs to slow climate change, share technology with less developed countries and cooperate in other undefined ways to reduce greenhouse gas emissions. It recognized that poor countries had a right to more economic development, which would involve more energy use.

In the stupefyingly slow crawl toward any eventual legally binding agreement, ratifying countries would send representatives to meet again and discuss specifics, in 1995, and once a year after that. But at the moment, the "UN Framework Convention on Climate Change" had about as much power to slow human induced global warming as the League of Nations had wielded before World War II, to stop even small wars.

The convention would take effect if 50 countries ratified it.

Having failed in his objectives, Bill Reilly was in a bad mood as he boarded a plane to Rio in early June. As head of the U.S. delegation, he would represent positions he detested, not only on climate change, but biodiversity too. Unlike every other major world leader, George Bush had decided not to sign the treaty to protect the planet's biodiversity, because in his opinion it failed to protect intellectual property rights, implied that biotechnology was unsafe and might involve expensive U.S. contributions.

"Rio was a tough affair. I said to my staff, guys, we've been given a very bad hand. We're going to play it as well as we can," Reilly said.

On the surface, at least, as tens of thousands of diplomats, journalists and NGO representatives flooded the city for the unprecedented meeting, a bright, happy time seemed to be beginning for the global environmental community. The Soviet Union had collapsed. The cold war had ended. The possibilities of attaining substantial progress in protecting the planet's environment seemed enormous.

"For the first time in centuries," wrote *New York Times* reporter Philip Shabecoff, "the planet was free of great-power competition for colonial expansion or military ascendancy. A new chapter in the long chronicle of geopolitics was clearly opening."

Thomas Pickering, U.S. ambassador to the United Nations, said, "This kind of opportunity comes along once in a thousand years."

Rio police had swept the city of its armies of homeless children. At Rio Centro, a gigantic modern conference center ten miles from downtown, up to 20,000 delegates, journalists and NGO representatives at a time swarmed through a complex including its own bars, restaurants, hospital and communications center. Brazilian marines guarded the center as UN Secretary General Boutros-Ghali opened the summit by asking for a symbolic "two minutes of silence throughout the world on behalf of the earth."

And silence is what Reilly probably would have preferred when his turn came to speak.

"I remember getting up to deliver the statement of the United States and thinking, this ought to be the high point of my career. I represent a country that has arguably achieved more on the environment than any country represented here. And it was just a very disappointing moment. After I sat down, a friend came over and said, 'You haven't lost it yet. Don't lose it now.' "

Conference organizers grew to share his frustration. For 12 busy days, for those who hoped the summit would achieve meaningful reform, like Tim Wirth or Al Gore, Rio remained a bright moment of opportunity. But by its end, even optimists described it ruefully as a "first step" toward some eventual muscular policy. The summit had fallen short of the grand hopes. It had achieved symbolism.

In the end, the 900-page blueprint for environmental action coming out of the conference, and called "Agenda 21," would not be legally binding.

Negotiators had failed to devise a treaty protecting the world's forests and had settled on a "Statement of Forest Principles," instead.

George Bush had flown to Rio, signed the toothless climate convention, and given a speech in which he said that the U.S. need apologize to no one because "its record of environmental protection was second to none."

Asked by a Brazilian reporter about criticism of the U.S. performance, Bush said, "I'm president of the U.S., not the world. I will do what I think is best for the U.S."

Before leaving, Bush attended a banquet for world leaders. Turning to greet the person to his left, he found himself looking into the face of Maumoon Abdul Gayoom, diminutive president of the tiny Republic of the Maldives.

The two men had never met. Gayoom as usual was worried about the big waves and high winds that had hit his country.

"We began to discuss environmental threats we were facing," Gayoom said. "I told him that if the U.S. and other advanced countries didn't take immediate action, a day might come when the Maldives would disappear. You know what he said?" Gayoom said, incredulously.

"He said, 'The United States will never let that happen.'

"I didn't take it seriously. I think it's something he said to make me feel a little safer," Gayoom added, smiling wistfully, going back to the disappointments of Rio in his mind.

———

But the framework convention on climate change, weak as it was, still remained a threat to those opposing energy cuts because later it could be filled in with targets and timetables. The fight was not over.

Both sides watched the skies. If the weather calmed, if the public came to believe that climate was "returning to normal," the calls for energy cutbacks would lose force.

Amid arguments over whether the record-breaking storms would stop, a hurricane expert named Mike Black strode across a Miami tarmac on the early afternoon of Saturday, August 22, 1992, toward a NOAA P-3 Orion four-engine turboprop plane. Black was a dark-haired, collegiate man who was fascinated by the killer storms, but whose academic view of them would change forever during the next two terrible days.

The Orion, which mechanics were preparing for flight into a hurricane, was rugged, big as a 727, and painted white, with an American flag on its tail, "Department of Commerce" on the fuselage and a cute depiction of the Muppet character Kermit the Frog.

Black wore long pants against the plane's air conditioning, a short-sleeved shirt, and he carried a jacket over his arm and a large black box filled with forms and aeronautical maps of the Caribbean and west Atlantic. He'd also packed "a rabbit foot for luck, and post-penetration popcorn, for when we're able to relax on the way back."

Black, 37, had flown into a dozen hurricanes by 1992, for the National Oceanographic and Atmospheric Administration's hurricane research division. It was his seventh season on the job. At the moment, a small and probably weak hurricane, "Andrew," was somewhere roughly 700 miles east of Florida.

Black's mission today was to help determine its track.

"There aren't any weather stations over the ocean," he said. "There's nothing over the ocean unless you go there. Also, in 1992, satellite data was *not* being injected into computer weather models to predict storm track. The computer power wasn't there."

To Florida residents, Andrew was not yet a significant cause for concern. In fact, the night before, as he left his office on Rickenbacker Causeway, Black had checked the forecast from the National Hurricane Center in Coral Gables and read the advisory "Enjoy Your Weekend."

The storm had been expected to turn north. But watching the Weather Channel this morning, learning that Andrew had *not* turned, he'd told his pregnant wife, "South Florida is not out of the woods yet. You may want to pick up some food and water."

"I was lackadaisical when it came to hurricanes. I'd seen many of them on my job, but I'd never had any in my neighborhood in South Dade County. Even knowing better, I had no shutters on the house. I had a minimal amount of water, batteries, supplies. I had no plans what to do if a big hurricane came. I'd always told my wife, if a category one comes, nothing to worry about. If a three comes, we'll talk about evacuating. If a four or five comes, we're getting you and the girls out of here."

The girls were Black's daughters, ages five and one and a half.

It was a beautiful hot summer day. Small, puffy white clouds drifted in the blue sky above sunbathers, packed outdoor cafes and strolling tourists.

As maintenance people ran through their preflight inspections, Mike entered the plane by ladder, ducking through a tail door. He instantly smelled "a unique smell that reminds me of burned electronics or ozone. Also, I know there's been a lot of vomit. They cleaned it up, but it's involved in the smell."

The crew today would be roughly 15 people, including 4 scientists, 2 pilots, a flight meteorologist and a flight engineer to monitor the engines, fuel and other mechanical functions. Over 50 navigation and scientific instruments were crammed into the plane, designed to measure the most sustained violent weather on the planet.

"You never know what you're going to get when you fly into a hurricane. You can be knocked out of your seat by turbulence, or it can be extremely mild. There have been nervous times when we lost engines. But I've never been concerned for my life. I feel it's safer than flying on a commercial airline."

Nevertheless, five hurricane hunter missions, Black knew, had never returned.

He would be radar scientist today, and took a seat at his work station, dead center in the plane, beside the left wing. Two computer screens in front of him would monitor the storm. "Belly radar" would provide a view like the one TV watchers see on weather reports. "Tail radar," Doppler radar, would provide vertical slices of Andrew.

"We'd be in the eye twice, to determine windspeed and central pressure. Most of the time we'd be on the periphery, dropping instruments to determine the storm's direction."

"Kermit" was scheduled to stay out as many as ten hours. It was heavy with fuel, and its slow, rumbling takeoff lifted them into the sky.

"It was gorgeous."

During the first hour, at 15,000 feet, Black chatted and walked around the aircraft, ate his steak-and-cheese sandwich and drank soda.

"There's not a lot of work to do in the first hour."

Then Andrew appeared on Mike's radar, 300 miles away, as an outer curving band of green and yellow rain.

Half an hour later the pilot saw cirrus clouds ahead, thin, wispy white streaks above the plane, at 45,000 feet.

The storm, on radar, seemed to move toward Mike.

"It was a little stronger than we expected."

Although Andrew was small, with only a few rainbands around it, the bands were clearly defined, meaning that the eye was closed. The hurricane was "well organized."

Like every hurricane hunter, Mike Black knew that each storm had a different personality. Some were persistent, refusing to die. Some were greedy, hitting every landmass possible. Some were unpredictable, especially the slow-moving ones.

"Right away I saw that Andrew could be a problem for South Florida."

Growing concerned, he went to the cockpit for a better view. It was growing darker outside although occasional hazy sun was visible sometimes through clouds. Bursts of light rain hit the Orion. The ocean, now that they were east of the Bahamas, was lightly choppy below, showing whitecaps.

The waves got higher and the whitecaps came more frequently. By looking at stock photos permanently kept on the plane, showing ocean conditions during hurricanes, researchers matched wave size to pictures, in order to approximate the force of the wind.

A voice in Mike's headphones said, "Looks like fifty knots to me. What do you think?"

"Yes. Fifty."

"At a hundred miles out we hit rainbands and mild turbulence. It felt like going through rain on your average commercial airliner while taking off or landing."

At 50 miles, the pilot's voice in Mike's headphones said, "Gentlemen, we're about to hit the eyewall."

The crew locked away flight bags, thermoses, clipboards or any other item that could become a missile. Mike buckled his seat belt but ignored the shoulder harness.

It grew *really* dark outside.

The plane pitched violently.

They were in the eyewall.

William Redfield, a Connecticut saddlemaker who could not afford his own horse, and therefore walked to customers, noticed something amazing in New England in 1823. Redfield, a weather buff, was strolling through the countryside after a hurricane and saw that trees and cornfields had been blown down to the northwest in one area, but a few miles later, the downed oaks and maples faced southeast instead.

That storm must have been a whirlwind, Redfield thought.

Until then, large storms were believed to be air masses moving straight across the earth, but Redfield published his new, correct theory in 1831.

By 1992, when Mike Black flew into Andrew, much more was known about the most deadly, long-lived storms on earth. Hurricanes can stir up over a million cubic miles of atmosphere every second. They can produce waves in deep oceans reaching 100 feet. Their coastal surges can send 20-foot-high walls of water to bury streets, cars, houses, and have killed over 100,000 victims at a time in Asia.

The destructive energy produced by heat condensation in just one hurricane, in one day, can equal the energy released by the fusion of four hundred 20-megaton bombs, according to NOAA. That's enough power, if converted to electricity, to supply energy across the U.S. for half a year.

And yet for all the brutality of hurricanes, they often begin gently, as Andrew did in 1992. Somewhere in West Africa around August 10 that year, Senegalese goat herders, tending their flocks, probably felt the briefest warm breeze waft across their faces as they gazed with pleasure into blue skies.

The breeze was part of the perpetual effort of the atmosphere to equalize pressure and temperature. It was caused by the contrast between the dry, hot Sahara, to the goat herders' west, and wetter jungles to the south.

"Air from the jungles flows under the air from the desert and that's where hurricanes start," said Hugh Willoughby, director of the Hurricane Research Division.

Passing over the goat herders, the atmospheric "wave" reached the African Coast, where it picked up energy from warm waters. Fishermen casting nets observed thunderstorms forming. A refreshing rain wet their bare chests.

"All summer, every four days, Africa spits out a little glob of vorticity," Willoughby said. "The vast majority never develop into hurricanes."

For the weak wave of air to actually grow into a mighty hurricane, conditions would have to be right. The ocean would have to be at least 80 degrees, to speed evaporation and help form thunderstorms. The surface winds would have to converge, pushing the thunderstorms into a cluster that can be as wide as 300 miles in diameter. Meanwhile other winds, higher up, would have to be moving in the *same* direction, pushing the storm cluster west instead of ripping it apart.

On August 14, those conditions existed. *Inside* the wave, wet air was rising, releasing humidity to fuel the growing storm. The condensing water cooled and released heat, and the heat caused winds in the thunderstorm cluster to become more powerful.

Still, Andrew was no more than an embryonic wrinkle in the otherwise uniformly flowing trade winds washing off West Africa. Watching satellite shots, the Tropical Analysis and Forecast branch staff of the National Hurricane Center, in Coral Gables, saw a gray-white frothy cottonlike image, an infrared shot of clouds passing over dark specks, the Cape Verde Islands.

It was far from a hurricane. But conditions worsened until on August 16 the forecast branch classified the thunderstorms as a "tropical depression," a cluster of storms around a center of low pressure, where circulating winds reach 20–38 miles an hour.

Other winds steered the floating power pack across the Atlantic, as the thunderstorms flared up or died down.

Two days later the storm was 800 miles east of the Lesser Antilles, appearing on satellite shots as vague curved bands encircling a white blob—colder cloud tops—in the middle.

As a threat, it was still not yet significant. It "went through terrible throes," said Frank Marks, another expert in the Hurricane Research Division. "It ran into a storm headed for Britain and almost fell apart."

Upper-atmosphere winds—blowing in different directions than the steering winds below—threatened to break up the storm, and the whole disorganized mass looked as if it would drift north over colder waters, lose its fuel supply and die. But instead it collided with a blocking high-pressure area sweeping down from New England. Since Andrew had to get out of the way, it shifted west, toward Florida, taking it over warm tropical waters again.

And the upper-level wind shear stopped.

By Friday, August 21, the day before Mike Black took off, weather satellites showed a small consolidated white blob with wispy curved bands of whiter gray around it—the storm's "outflow," air shooting out of the heat engine like exhaust blowing from a chimney funnel 40,000 feet high.

Inside the storm, intensifying evaporation was causing condensing warm air to spread out in all directions as rainbands.

Back in Coral Gables, the National Hurricane Center upgraded Andrew to tropical storm status as winds reached 39 miles an hour. Then, just before Mike Black took off, the winds reached 74 miles an hour. A small pinhole eye appeared on satellite shots.

Andrew had become a hurricane.

"We were getting concerned," Black said.

But which way would Andrew go? How strong would it get? Based on its current strength, Andrew was assigned a number, one, under a classification system devised originally by a Coral Gables consulting engineer, Herbert Saffir, and the director of the National Hurricane Center at the time, Robert Simpson.

Under the system, a category 1 hurricane was strong enough to put mobile homes and power lines at risk, but average homes would withstand most damage.

If Andrew grew into a class 2 hurricane, its winds of between 96 and 110 miles an hour would be able to smash road signs, snap tree branches and put mobile homes at more risk.

If winds reached 111 miles an hour, Andrew would become a class 3 hurricane, lifting roofs, breaking walls off buildings and uprooting trees.

But class 3 was minor compared to class 4, where winds as high as 155 miles an hour could wreak enormous structural damage and send killing storm surges onto the state's low-lying beach communities.

The nightmare category, however, was 5, where catastrophic winds could level whole communities.

On August 22, the irony for Mike Black, Frank Marks and other meteorologists they worked with was that for all their knowledge of hurricanes, many of them had not prepared for this one. Like millions of other U.S. residents who had flooded into coastal communities during the 1970s to 1990s, they'd not personally experienced a serious hurricane yet.

"For years, we guys in the lab had talked about what would happen to our homes if we ever had to fly if a hurricane hit Miami," said Frank Marks, originally from Westchester, New York. "Unfortunately all we'd done is talk. We'd never arranged for a crew of volunteers to fly, so other people could make sure their families or homes were prepared for a storm."

Stanley Goldenberg, another meteorologist, originally of St. Louis, had a wife due to deliver their fourth child on August 23, and was not flying into the storm because of the impending birth. But he hadn't prepared for the hurricane either.

"I did not have a bunch of money to rush out and buy plywood. I didn't have shutters made. I thought, I can't spend that kind of money, and I don't have the time or feel like hammering it in, anyway."

He was on the phone, though, by the time Mike Black flew into Hurricane Andrew, calling in-laws and friends, warning them, "The hurricane might be coming. You won't hear it on the news yet. They probably won't issue warnings till Saturday night or Sunday morning. It could turn away. It could fall apart, but it could intensify until we have a very serious situation on our hands."

———

As Mike Black's plane hit Andrew's eyewall, Black felt the plane rock and saw the wings "flapping around."

"Your only reference point is the aircraft itself. Visibility stops at the wings."

The turbulence made his instruments look blurry.

"I love going through eyewalls. A lot of people get sick and can't stand it. To me it's like being on an amusement park ride."

After four minutes, the turbulence stopped and it grew brighter outside. Then Black saw a curtain of rain, and they passed through it into the eye, blue sky spotted with small clouds.

"It was like being in an enormous stadium, twenty miles across."

The plane banked, beginning an orbit around the eye. Mike looked up at white masses of rotating clouds, boiling sides of the bulging eyewall.

Looking down, he saw, in the sea, "Thirty- to ninety-foot waves. The storm was a huge mixmaster. In the eye waves were crashing into each other from all directions."

Scientists on board regularly shot "dropsondes"—wind measuring devices—into the storm, inserting the meter-long cylinders into transparent launch tubes in the aircraft's floor. Each time the flight director would say, "Three-two-one," an engineer would pull a trigger, and with a whistling sound, cabin pressure would blow the devices out of the plane.

The crew tapped at keyboards and Black studied his radar. The plane stayed in the eye only a short while, then punched back through the wall and began a steady check of Andrew's strengthening perimeter. The dropsondes showed the winds increasing. It was clear to all on board that the storm might present a big danger to South Florida.

"We were getting anxious.

"We got back about ten-thirty, and I went into the NOAA Aircraft Operation Center, gave a synopsis of the trip and asked what was happening. The storm was still intensifying. Miami was under a hurricane watch. A hurricane warning would probably be issued by five A.M. I thought, I *gotta get home*."

Once there, he and his wife, Sandra, loaded up their '87 Honda and she headed off for Tarpon Springs, to Mike's parents' house, with the girls. The normally five-hour drive took double that, with Mike behind her part of the way in a government van. He'd been dispatched to Tampa to wait for the hurricane and measure it when it crossed to the west coast of Florida.

Reaching Tampa, he checked into a motel, turned on the TV and waited, sick with frustration.

The storm was headed toward his home.

———

All over Miami, residents hammered plywood sheets over windows. They flooded supermarkets, buying supplies. They created gigantic traffic jams, 200 miles long, on Interstate 95 as they tried to evacuate. Florida trooper Pam Maney, watching the stupefying parade of evacuees, said, "Nobody around here has ever seen anything like this."

Tolls were suspended on roadways. Florida Power & Light shut down their four nuclear reactors. Rental car agencies emptied out. In Orlando, Walt Disney World's 15,739 rooms were all booked by people seeking to flee the coast. Heavy coast guard vessels were sent out to sea. Smaller ones were secured onshore.

"Andrew's on a dead course for South Florida," Bob Sheets, director of the National Hurricane Center, told reporters in the building as the eye headed directly at headquarters. "I hoped I would never experience this."

———

By now, the meteorologists at the Hurricane Research Center had realized that this time they weren't going to be studying a storm that struck someone else's city. This time they'd be at ground zero when the damn thing hit.

Frank Marks spent Sunday morning hammering plywood over the windows of his rented townhouse in Sabal Chase, while Andrew raked the Bahamas with 120-mile-an-hour winds, and gusts up to 150 miles an hour. Marks designated a family room with only one window as a "safe room." The family hauled valuables, food, mattresses, a gas grill to brace doors, sleeping bags and camping lanterns into the room.

"I was on edge. My wife said I was pacing like an animal. Was this door locked? Was this window secure? Did I forget something?"

As he worked, he got a call from a French hurricane expert staying at a nearby hotel that was closing because of the storm. Marks invited the Frenchman and his wife over. The couple had never been in a real hurricane and they were excited. They brought wine.

By late afternoon it was "eerie" but "comforting" in the living room, with the TV on, and six people: Marks, his wife and daughter, a nephew and the French couple watching news of the coming storm.

But suddenly Marks realized he'd made an error. The "safe room" where everyone was supposed to retreat in an emergency *lacked a door*.

"I decided the emergency room would now be a half bathroom downstairs."

It was a tiny room, with a vanity and a toilet. He put candles in the room, and matches. Hurricane Andrew, roaring down on Florida, had become a category 4 storm.

He drilled the kids. "If I say '*move*,' don't ask questions. Run."

Stanley Goldenberg's wife had gone into labor at 9:00 A.M. Sunday, possibly induced early by low-pressure air from the approaching hurricane. Goldenberg stayed with her throughout the successful birth, and then headed home.

He found his brother and sister-in-law at his house, with their three boys. They were preparing for the storm.

"We covered up every window except one small one, hammered on the outside with mason nails, not the best way to do it, but it was the best way we had. We buckled down the house. My sister-in-law filled containers with water, and stored away all my tax papers in a closet.

"Unfortunately we had no interior windowless room to pick as a hurricane safe room. So we were prepared to get everyone into the hallway, covered with blankets against flying glass, if the need arose.

"By nightfall we hunkered down. I was thinking, it's *so* nice to be prepared. We're gonna be okay."

He was wrong.

"Everyone conked out. I slept a few hours, got up about two-thirty A.M. The electricity was still on. It went off around four. The wind started to pick up. I woke everyone, said, you gotta be awake now. I got everybody out of their beds and into the living room. I knew from the TV that sometime around four-thirty, the eyewall would go through."

The assessment was close. At 4:35, radar at the National Weather Service's Miami facility made its final sweep before one of Andrew's 164-mile-an-hour gusts wiped it away. The terrifying picture showed watchers in the shaking building a purplish doughnut shape, pulsating with power, almost surrounded by green and yellow rainbands as it crossed Elliot Key and made landfall in an arc sweeping from south of Key Biscayne to north of Key Largo. Turkey Point's nuclear reactor was covered in red. Homestead was about to be overwhelmed by red. Florida City lay in the path of the coming storm.

The eyewall hit Stan Goldenberg's house.

"Things went crazy. The first thing that happened was that it ripped off the living room window," said the slight, fast-talking, usually grinning meteorologist, a religious man, and a man whose politics leaned to the right.

"In the plane we get bumped around a little when we fly into hurricanes. On the ground it was worse. In the plane we go in quick, through the worst part of the storm. But here it came on gradually, and now it was hitting a house not made for it. We had six boys in there, ranging in age from two to ten. We had a small kitten and three adults in the hallway, under the covers.

"The sound got louder. Wind started ripping through the house. We heard other windows breaking.

"My brother-in-law suggested that we get to the garage. His station wagon was there and he had faith in that station wagon. It was like an armored truck.

"So we crossed the house as the wind whipped through it. My four-year-old tripped and the wind started pulling him out back, but one of the other kids grabbed him."

Reaching the garage, a shocked Goldenberg saw that part of the wall was gone, and the wind inside was so fierce the family backed into the kitchen, crouching down in the howling wind. Some of them didn't even wear shoes, but sandals, flip-flops, and there was broken glass all over the place. With no place to go, they huddled on the floor, between counters, covering their heads.

"My sister-in-law had filled pots with water and now the pots started falling on us.

"I covered my kids. We were praying. I don't say that lightly. I am a believer in the Lord and we pray all the time. But we were praying more than normally. We were crying out to God. Actually, screaming would be a better word.

"Then something very heavy fell on me, and I heard my brother-in-law say, 'The roof is gone.' "

What had hit him, and pinned him, was a kitchen wall whose fall had been arrested by the kitchen counter.

"I didn't know if I could move. I was on my knees crouched over the kids, with my foot extended, and this weight on me. I was in excruciating pain. My son was beneath me, his head on the ground. The water level was rising from the rain. I would check his head regularly. I could have pressed it into the water and suffocated him.

"Everybody was screaming. One of my nephews was screaming, 'We're gonna die! We're gonna die!'

"I couldn't take it anymore. I felt like I was suffocating. I was under the comforter that had gotten all wet, and it was a very close space, and the force of the wind made it seem like it was compressing more on us."

At length, Stan realized from the sound of the wind that it had switched around. "We were at a changing point. The winds were a little calmer. I realized we better move now.

"I told everyone, it's going to get stronger again when the back side of the storm hits.

"As the wind abated we were able to move the wall off me a bit. They got me out. I could stand but there would be nerve damage. We headed for the garage again. The house seemed totally demolished. The kitten had blown away, into a pile of rubble, but we pulled him out.

"When we got to the garage . . . well . . . only one wall was standing . . . we got in the car and put the carpet over our heads in case the windows smashed. I began to think, *Should we stay here?* The hurricane could easily crush the car. Should we try to go across the street to the neighbor's house, which was standing? But if we got over there and knocked, they might not be able to hear us."

As the family sat stunned, wet and unsure what to do, they were astounded to see two figures running toward them through the storm.

"Do you want to come to our house?" asked the neighbor and his wife.

Everyone in the car cried out, "Yes!"

———

Mike Black, in a motel room in Tampa, needed to find out if his home had survived. He could not go back yet, even after the storm had passed through Miami. He was still working.

"It was odd sitting in a motel watching TV, seeing landmarks like Bloomingdale's with the whole front pulled off. Bloomingdale's is ten blocks from my house."

Mike called friends and asked them to drive to his house, and see if it had survived the hurricane.

"They couldn't *find* the house at first."

All the landmarks were gone. There were no street signs. There were no telephone poles. But when they did locate the address they called Black and said, "Want to hear the good news or the bad news?"

"Just tell me."

"The front looks pretty good, but the back, uh, isn't there."

———————

Frank Marks's home lost only one window in damage, but Stan Goldenberg's house was destroyed, along with 126,000 Florida homes, demolished or damaged by winds that some scientists believed had reached 200 miles an hour.

In its brief violent assault on South Florida and Louisiana, Hurricane Andrew had killed 44 and inflicted more damage than any natural disaster in U.S. history.

Initial estimates topped $25 billion in Florida, with U.S. insurance companies predicting that by 1995, the total economic impact would rise another $10 billion. One hundred sixty thousand people were homeless in just Dade County. The municipal power grid in Homestead and Florida City was destroyed. Homestead Air Force Base was leveled. Banking had come to a halt, and with it, said a Department of Commerce report, "much of society's ability to function."

But had global warming been involved, in any way at all, in Andrew's life?

Mike Black couldn't decide. At first he felt that climate change had not affected Andrew. "It was the third strongest hurricane to hit Florida; a normal once-in-a-lifetime event."

But by the end of the decade Black would be wondering about the increase in category 4 hurricanes.

"There is a change," he'd say. "But is it long term and related to global warming, or a shorter, decadal change?" He'd frown. "Global warming or natural cycle? I don't know."

Stan Goldenberg would scoff at any suggested links between the storms and the greenhouse effect. The number and severity of hurricanes had always fluctuated, he felt. Some years brought more and stronger storms and some years brought less.

Hugh Willoughby would say, "If global warming is coming, we'll have a problem with storm surge, and we may have an increase in the most dangerous storms. The one percent of hurricanes at the tail end of the distribution are going to be meaner . . . The worst-case scenario won't be category four anymore. We will have to concern ourselves with more than once-in-a-century category fives making landfall on the United States."

And as for Tom Karl, up in Asheville, working on the next IPCC report, it was simply still too early to make statements linking global warming with hurricanes. He had to look at the big picture. The whole world. Believers in global warming might warn that immediate action was needed to curb it, but for Karl, Hurricane Andrew's high winds and storm surge went into the larger mix of material to be considered by the IPCC.

Andrew was just one more storm in the blend, and no single weather event could be linked in any way to climate change. Not to Tom Karl. Not yet.

To Karl the careful statistician, as 1992 ended, more work was needed before scientists had the kind of answers that George Bush or John Sununu were waiting for.

But Jim Hansen, in New York, was smiling by Christmas. As his climate model had predicted, by the end of 1992, global temperatures were going down.

Washington ... Missouri

KATHLEEN MCGINTY, new young appointeee at the White House, was getting frustrating lesson after lesson in the limitations of power in the spring of 1993. As head of Bill Clinton's White House Office on Environmental Policy, the 29-year-old was charged with initiating and coordinating every major battle, bill, proposal and strategy to institute the president's environmental plans across the U.S.

Whether the problem involved climate change or saving Alaskan forests, McGinty stood between the White House and Congress, environmentalists, Clinton's own cabinet, and industry.

In contentious meeting after meeting, she was having the kind of squabbles she never dreamed even existed when she was growing up, one of ten children of a Philadelphia policeman, listening to her mother constantly point out beloved birds and trees as she formed the child's environmental attitudes.

"I'd never even heard of Al Gore four years earlier."

With George Bush defeated, the new occupant of the White House had promised to end a recession and elevate environmental issues to top priority along the way. Suddenly Al Gore, Tim Wirth and Rafe Pomerance were inside the administration instead of sniping at it from outside. Their commitment to achieving greenhouse gas emission changes would now be publicly tested—especially since the climate convention

Bush had signed in Rio, weak as it was, obligated the U.S. to design *some* kind of action plan.

But Clinton's problem—and McGinty's—was, what if policies to mitigate greenhouse gasses made the fumbling economy *worse?*

McGinty was a tall, lean woman with blue eyes, high cheekbones, pale freckled skin and long auburn hair, which she tended to toss. She'd received a BS in chemistry from Saint Joseph's University, and a law degree from Columbia University in 1988. She'd worked as a research assistant for Atlantic Richfield Chemical Company, specializing in waste water treatment systems, not climate.

"It was a coincidence I ended up with Al Gore."

Her meteoric Washington rise had begun in 1989, when she won a one-year fellowship from the American Chemical Society to work on Capitol Hill and was assigned a position on Senator Gore's staff.

Gore found her smart, tough and a quick study.

"After my fellowship was over he asked me to come aboard as environmental policy adviser. I did not come as an environmentalist . . . I knew nothing about the intricacies of environmental policy, but I brought a strong passion to the issue because of a love of the outdoors."

With Al Gore, McGinty had toured the country and seen up close many endangered trees, birds, glaciers and forests. She'd worked on Gore's climate change hearings, and on coordinating greenhouse strategy with Tim Wirth, whose office had been down the hall, and with Rafe Pomerance, whose big, excitable form had been a familiar presence rushing down the Senate corridors. She'd worked on issues including deforestation, the ozone layer and even environmental consequences of foreign trade.

But in 1993 she was learning the difference between pushing issues in the Senate and pushing them in the White House.

"My development went through initial growing pains. When we were in the Senate we were an operation of one: myself and Al Gore. In ten minutes we could be forcefully articulating a position . . . But it's another thing to craft action that will deliver results. You can't just dictate things. You have to involve a variety of players so your policy will be better in the end and people *do* it."

The problem was, at the start in the new administration, the "variety of players" disagreed about climate just as their predecessors had during the Bush years. The White House had changed but the fundamental competing interests—economic versus environmental—had not. Gore urged Clinton to commit to freezing greenhouse gasses at 1990 levels by 2000. Treasury Secretary Lloyd Bensten argued that more study was needed on the potential damage to industry if carbon dioxide emissions were curtailed.

What *had* changed at the top was that Clinton and Gore were actively seeking a policy to curb greenhouse gasses. They did not doubt the science. But balancing a futuristic climate disaster with a current economic one was delicate, and Clinton was unsure how to proceed.

In meeting after meeting, McGinty found herself enmeshed in maddeningly slow talks.

"You wish a meeting that was four hours long was forty-five minutes long. And at the end of four hours you're back where you were at the beginning."

Nor was the issue even close to being the top problem facing the new president. By spring 1993, Clinton was on the ropes, fighting the hugest battle of his new administration, trying to push through a budget and tax hike to help curb the country's crippling deficit.

Only later would administration officials realize that the outcome of that fight would determine much of their global-warming policy for the next seven years.

The reason was, a section of Clinton's budget proposal already *included* a way of dealing with greenhouse gasses—a new tax that would curb them while it cut the deficit. The tax would be a fundamental change in the way the government waged energy levies, and it terrified coal and oil companies.

The "Btu tax" proposal was causing a huge fight.

The controversial tax would levy the amount of heat energy released by different fuels. A Btu, or "British thermal unit," is the amount of energy needed to raise the temperature of a pound of water by one degree. Since different fuels emit heat at different efficiency levels, the tax would require energy companies to pay more if they used dirtier forms of energy, like coal or oil, less for cleaner energy, like nuclear, and nothing for solar or wind energy.

The tax would raise $70 billion in five years to combat the deficit and offer incentives to switch to clean energy along the way, proponents said.

"If Congress passed the Btu tax we would be instituting the Rio Climate Convention right out of the box," said Rafe Pomerance, who was now an Assistant Deputy Secretary of State for Environmental Development.

But opposition had reached fanatical heights. Oil companies poured hundreds of thousands of dollars into anti-Btu-tax-campaigns. The American Petroleum Institute warned that Clinton and Gore wanted to "move the nation away from oil and toward a dramatic decrease in average energy use." "Citizens for a Sound Economy," a Republican group, paid for huge ads in Oklahoma—home of key senator David Boren, who sat on the Senate Finance Committee. The ads said Btu stood for "Big Time Unemployment."

Boren supported the budget proposal at first but changed his mind.

"The climate policy of the administration, in retrospect, was determined by what happened to the Btu tax," Rafe Pomerance said. "It cast a long shadow."

By spring, "the tax was not handled well," Pomerance said. Clinton offered so many loopholes to opponents, trying to entice them to support the tax, that critics laughingly called it as porous as Swiss cheese. The Senate balked, led by Senator Boren. Clinton withdrew the proposal in order to try to get the bigger deficit reduction package passed.

"The loss of that tax was the key turning point on environmental policy, because you can't control CO_2 without energy pricing," Rafe Pomerance said. "The withdrawal limited our ability to do anything domestically . . . Once the tax failed, we could never return to something that strong, never return to a pricing instrument."

And so, only months into the new administration, Katie McGinty found herself coordinating an effort to devise an official plan to reduce U.S. greenhouse gasses, but how strong could such a plan be if a Btu defeat tied White House hands?

By April, with the budget fight still ongoing, as Clinton prepared to commemorate his first Earth Day in office with a much anticipated speech on the environment, "None of the ideas the White House is considering for inclusion in Mr. Clinton's speech have caused as much

disagreement as the global-warming proposal," the *New York Times* edi-
torialized. A commitment to reduce the levels of carbon dioxide that
the United States puts into the atmosphere would affect every American
and put more pressure on American industry to compete with Japanese
and German manufacturers who already use energy more efficiently."

A top aide at the Department of Energy tried to soothe reporters
who sensed the president might be backpedaling. "It may take more
time to work this out," he said.

Which was what Katie McGinty was trying to do.

"We started with a blank slate in 1993," she said. "We engaged in an
exhaustive process on the issue from the first days in office, until the
president's speech . . . to put him in a position to . . . sign the U.S. up to
a goal of trying to return to a 1990 level of emissions by 2000 . . .

"First the scientists had to tell us what we know about climate
change. The Department of Commerce had scientists who had been
working on the issue. The Department of the Treasury had economists
who had been studying it. The EPA had scientists who could talk about
the health impacts from particular pollutants. The Department of
Energy had a whole bank of atmospheric scientists who could talk
about available technologies that could be deployed to reduce the
gasses . . . A vast army of talent within the government."

It sounded from her description like the scientists and economists
were a finely attuned team marching toward some easy-to-reach policy.
But the truth was that different departments were producing different
conclusions. Treasury Department studies predicted that mandatory
energy cuts would damage the economy. EPA studies concluded that
cuts could be accomplished painlessly.

Meanwhile, "someone like Larry Summers at the Treasury Department, distrusted among the ecological crowd, wanted to see the numbers," McGinty said. "He'd say, what do we know about the feasibility
of reaching these targets? How does it break down among industrial
sectors? Are the actions we can take realistic? Are they actions where the
president has authority on his own, or does he have to get things passed
by Congress? What economic impacts are we talking about?"

The answer depended on who was asked.

On April 21, Earth Day eve, Bill Clinton stepped up to a podium in the Botanical Gardens, in the shadow of the Capitol, and addressed a rapt audience of environmentalists who were eager to hear what the president they had supported would propose.

They liked when Clinton vowed to commit the U.S. to "reducing our emissions of greenhouse gasses to their 1990 levels by the year 2000."

They were less enthusiastic when he didn't say *how* such a thing could happen, but he promised that a specific plan would be ready by August.

When the speech was over, Katie McGinty, trying to get the plan in shape, went back to work.

———

The change started with a small southward alteration of Pacific current and a new wind carrying the squawking seabirds over the ocean off Peru, as warm water displaced the normal cooler upwelling in the Eastern Pacific. Fishermen shook their heads, because with the cool waters gone, tuna and yellowjack were leaving. Peasants on land looked up as skies darkened, and they crossed themselves because they knew what was coming. Then the fierce rains began, and the mud slides did too.

The pattern was older than the Roman Empire but the severity was about to get worse. This cyclical alteration in ocean currents usually occurs twice a decade, usually lasts about 18 months. Peruvians call it El Niño, "boy child," after Christ, whose birthday often marks the occurrence. In the 1920s, scientists believed the phenomenon limited itself to the west coast of South America, but in 1957 they discovered that Pacific waters in El Niño years warm as far away as the International Date Line, a quarter of the way around the globe.

By the 1970s, the scientists realized that El Niño was part of a gigantic coupling of the ocean-atmospheric system that could affect weather thousands of miles from Peru. A severe El Niño could send record rains to pound Louisiana, or withhold rain as a burning sun dried out Mexico, Australia, Indonesia. It could spin vicious hurricanes across Polynesia, snapping trees, shattering homes.

El Niño was one more part of the global climate cycle. And like other parts it was driven by heat. If the heat coming into the ocean and atmosphere intensified for some reason, El Niño's effects—like the rest of the atmosphere's—would grow more severe.

By 1992, around the time Bill Clinton had asked Al Gore to join his political ticket, the latest El Niño, which had begun in '91, had been forecast to end. But by the time Katie McGinty started working on the administration's climate plan, the El Niño had grown oddly stronger.

In fact, 1993's El Niño was becoming the longest on record, said Gerald Bell, a meteorologist at NOAA's Climate Analysis Center in Washington.

"It's unlike anything we've seen," agreed Donald Hansen, an oceanographer at the University of Miami. "Maybe it's not just an El Niño, but it is signaling that two or three years ago we had a change in the climate."

Already in 1993, the "prolonged El Niño event," as a UN World Meteorological Organization climate report called it, had caused "significant precipitation anomalies," including a "devastating drought in southeast Africa, the worst in 100 years," and a "vast belt of copious rain which stretched from Oman over Iran, Pakistan to Sinkiang and in southeastern South America over Uruguay, neighboring parts of Argentina, and southern Brazil."

To make matters worse, "Pacific temperatures, already above normal, are increasing again," Dr. Bell warned reporters. "We're putting out the red flags now."

In the U.S., the bizarre El Niño was about to add to Katie McGinty's problems, cause Missouri governor Mel Carnahan his first big crisis in office and threaten Vern Bauman's life.

The alarming progression of effects began when water temperatures near the International Date Line, normally 83 degrees Fahrenheit, rose to 85 degrees and kept going up. Then prevailing winds blew the warm Pacific water east and the excess heat caused immense thunderstorms to erupt.

The thunderstorms disrupted the jet stream—a high-speed river of air—as far away as the northern hemisphere, meteorologists said, watching as the jet stream in the U.S. detoured around a "freakish" high-pressure system over the Southeast and locked itself above the Midwest.

Since storms tend to break out along the jet stream, far below, Governor Mel Carnahan stood watching rain begin to fall at the window of his spacious, domed Jefferson City office. A quiet, self-contained man of average size, he had been in office only six months, like Bill Clinton. The furniture was all antique. A dark blue carpet covered the marble floor, and showed the state seal in the center. At the governor's back, oak-paneled walls featured murals commemorating famous Missourians whose achievements had made history, as would the incoming storm.

Out the window, and past the Capitol parking area and its pear, poplar and evergreen trees, Carnahan gazed at the soon-to-be-swollen Missouri River below the Capitol bluff.

About the same time, along the eastern border of the state, a road construction company owner and local Mississippi River levee official named Vern Bauman tilted his hardhat back and eyed, from a high dirt levee, the brown turgid swatch of Mississippi River separating his town, Ste. Genevieve, from Illinois on the far bank. Rain pummeled the barges, ferries and river tugs. Ste. Genevieve was one of the oldest settlements on the waterway. Rain battered its levees and the railroad tracks that paralleled the river, the soybean fields abutting the river and the fine old remodeled tourist homes that served thousands of visitors each year.

Bauman saw levees lining the Illinois side too.

When the rains didn't stop, after some days, Missouri's rivers started rising.

But Missourians are used to rising rivers. In Jefferson City and Ste. Genevieve, Carnahan and Bauman grew concerned but not worried.

Meanwhile, after a wet winter, the ground throughout a nine-state area of the Midwest—a gigantic natural bowl draining into the Mississippi and Missouri Rivers—was growing saturated. Hundreds of thousands of square miles of runoff kept washing into the rivers.

A NOAA "Natural Disaster Survey Report" would call the unabating rain the "direct result of major global-scale anomalies which can be attributed to significant climate variations."

Mel Carnahan and Vern Bauman got worried.

In early July, a TWA jet banked toward Rome International Airport, through blue sky, toward the Colosseum, the Vatican and the historic city below. Buckling his seat belt, Mel Carnahan felt some of the ease a man should experience when taking a vacation, but he could not rid himself of an uneasy premonition. He'd been promising his wife the Rome trip for years, but one problem or another had always delayed it. First there had been legal cases to handle, and then a political campaign.

Right up to his departure, he'd been undecided whether to go because of the floods back home.

By July 3, a week earlier, Carnahan had been dealing with the rain for months. In the worst flooding since 1965, the Mississippi River had overrun thousands of acres of farmland in Missouri, Minnesota, Wisconsin, Illinois and Iowa. From St. Louis to St. Paul, a 500-mile-long stretch of river had been closed to traffic.

The Mississippi had reached 12 feet over flood level in Hannibal, Missouri, the highest it had risen in 20 years. But then the sun had come out for the 38th annual National Tom Sawyer Days, blue sky had blanketed the state and spokesmen for the Army Corps of Engineers had predicted that the river would drop below flood level within two weeks.

Still unsure whether to go, Carnahan had toured flood areas on July 6. From a National Guard helicopter, he'd looked down on flooded farmland, river towns busily sandbagging and even convict labor he'd approved helping shore up defenses. Logbooks at SEMA, the State Emergency Management Agency office, showed that local officials across the state were reporting that conditions, while threatening, were under control.

As Carnahan interpreted things, "We'd had a lot of spring rain, and some typical flooding in June and early July . . . We were really past the season. There was quite a lot of water but I thought we were on the subsiding side of things."

But now, arriving in Rome, the governor called home to hear the sort of news dreaded by vacationing politicians.

"The flooding was increasing to serious proportions."

Talk about quick vacations.

"We stayed in Rome nineteen hours. We left the next morning. We toured the Sistine Chapel. We had an hour or two there."

Carnahan was fairly popular back home at the moment, having been elected by a landslide, with over 1,300,000 voters expressing confidence in his ability to steer Missouri through calm times and emergencies. But during his two-day absence, an emergency had been declared by President Clinton at the request of Missouri's lieutenant governor.

With a frustrated Carnahan on his way home, more rain pounded the Midwest. The Mississippi River reached 22.5 feet at Davenport, Iowa, breaking the previous record. Eighty percent of Missouri's 8.5 million acres of prime farmland was underwater and three thousand homes had been evacuated since spring. Illinois, Wisconsin, Iowa and South Dakota officials reported crop losses of over $1 billion so far.

Mel Carnahan was not a visibly demonstrative man. "His emotional band was very narrow," his aide, Roy Temple, said. "Normal and very upset could be indistinguishable to the layperson."

Carnahan tended not to discuss issues in emotional terms.

"A shake of his head was pretty profound for him."

But Carnahan could envision what was going on back home. Levees were being smashed, further flooding farmland and towns. Uprooted trees were torpedoing into bridge buttresses already scoured by silt-filled waters. Volunteers and convicts working side by side were getting lessons on how to fill sandbags. ("Halfway. Then fold 'em.") National Guard troops were rumbling into riverfront towns.

Missouri had the most river frontage of any midwestern state, and the misfortune not only to receive the historic rain falling on it but runoff surging toward it from one of the hugest natural drainage systems on earth: that of the combined Upper Mississippi and Missouri Rivers. Just the northern Mississippi alone, flowing down from its headwaters in Minnesota, was swollen with runoff from the Red River, Minnesota River, Iowa River, Illinois River.

And the Missouri River, extending back 2,460 miles from its beginning at the confluence of the Gallatin, Madison and Jefferson Rivers in Montana, drained ten states, 529,350 square miles. Engorged also by the North Platte River, Kansas River and Big Sioux River, it flowed past Kansas City and across the length of Missouri, to meet the Mississippi near St. Louis.

Now, because of the unprecedented weather, the governor faced the kind of problem more and more public officials will struggle with in the very near future if IPCC predictions are right—the logistics of dealing with weather disaster. Now Carnahan's political skills would be tested as much as his citizens' ability to pile sandbags. He might be satisfied with the way his emergency services department handled difficulties inside the state, but the department had no power over another problem—one which will be devastating in third world countries, but still formidable in developed ones.

Missouri needed *money,* lots of it, fast.

Carnahan would have to ask Bill Clinton for help. He might be the top official in Missouri, but in Washington he would be just one more supplicant holding his hand out for cash. *All* the governors in the nine-state flood area would need money. They were *all* flying around their states, looking down in dismay as their staffs tallied damage in what would be the most costly flood in U.S. history.

And all the governors knew, as Carnahan did, that in a recession, neither the White House nor Congress would be in a mood to hand out the kind of money they'd want.

At least Carnahan's background had prepared him a bit for lobbying in Washington. His father had been a U.S. congressman for 14 years before President Kennedy named him first U.S. ambassador to the African nation of Sierra Leone.

So Carnahan knew the Capitol somewhat.

His political skills had been honed during a steady climb inside the state. After earning a bachelor's degree in business administration from George Washington University, joining the U.S. Air Force during the Korean War, and graduating with a law degree from the University of Missouri, he'd been in politics by age 26, elected municipal judge in his hometown of Rolla. Two years later he'd moved up to the state house of representatives. In 1980 he'd been elected state treasurer, in 1988, lieutenant governor.

When reporters asked him about his motivation for seeking public office, he told them about Adlai Stevenson the Second, who called public service a "high calling."

Now Carnahan would get a chance to succeed or fail at his high calling—because of a storm.

Flying home from Rome, he awaited his first sight of Missouri, which came as the TWA jet reached St. Louis and he found himself looking down on a swollen waterway rampaging south, much of it already out of its channel.

Tens of thousands of people down there were piling sandbags. None were thinking about El Niño, carbon dioxide or climate change. They were trying to save their homes.

Carnahan needed to gauge the damage as quickly as possible, and soon he was in the air again, in a National Guard helicopter, flying over the floods.

He went up the Mississippi to Canton, almost to the Iowa border. "It was very dramatic. They were putting stakes up, and ordinary sheet plastic and sandbags behind it. They were extending the levees back a little higher. Alexandria had been flooded. Canton was fighting it off."

He flew over Hardin, Missouri, which would soon experience the most gruesome episode of the flooding.

He flew toward the town of Louisiana, but heard that its mayor was threatening to shoot down the helicopter.

"The mayor was under stress . . . The water got within six inches of the top of their levee. If they flooded, the whole town would have been wiped out. And the tops of these dikes were homemade. It's like a chain with the weakest link. If one place went, the whole thing would go. Even the motion of the helicopter blades was very tenuous. That mayor popped off to the security folks that if we fly over his town he's going to get out his gun.

"We obviously did not fly over his town."

The disaster grew worse daily. Tornadoes did their primary damage in minutes. Hurricanes generally passed in a day. But the floods were not stopping. The rains would cease for a while and the sun would come out, but then clouds would gather again overhead, or to the north, and the downpours would resume. The trickle of runoff would grow to streams. The streams would become deluges.

One day Carnahan found himself in St. Charles, on the Missouri River, speaking to a man who talked about being backed up on his bed, in his house, killing water moccasins trying to get out of the water, onto his bed.

"The governor just couldn't comprehend how people had to do the kinds of things people had to do to survive," Roy Temple said. "He was not an emotive person but he always wanted to talk about it."

Mayors were calling him, needing equipment, National Guard help, housing. "I could tell if the local jurisdiction had a competent director by the questions they asked."

One question that made Carnahan doubt the competency of the caller had come from the mayor of St. Louis, who woke him up in the middle of the night to say the city faced an emergency at a Phillips Petroleum Pipeline Company propane tank farm.

"They were flooding. They hadn't gotten the tanks out of the way."

The mayor feared that the 51 tanks, each holding 30,000 gallons of the highly explosive propane, would be dislodged by rising waters and turn into torpedoes directed toward indiscriminate targets by the floods.

Already a few tanks had started leaking. If the propane ignited, the combined explosion would form a fireball over south St. Louis one mile wide.

"They didn't have plan one what to do about the tanks . . . Nobody ever thought of this . . . he calls me at one in the morning saying, '*what do we do?*'!" Carnahan said.

Police diverted traffic and helped evacuate the area. Professional divers disconnected feeder pipes from the tanks, to keep the pipes from breaking. And small amounts of propane were cleared from the pipes.

Disaster was averted.

As the days went by, "I was consumed by this. We kept thinking the rains would stop."

Time after time the Army Corps of Engineers would predict a crest, a flood high point, and the crest would pass, which was supposed to mean the worst was over, but afterwards, astonished Missourians would learn that the corps was forecasting a *higher* crest.

"We've never seen anything like this," a spokesman for the corps said in Rock Island, Illinois. The rainfall was breaking records kept since the early 1800s. Some critics would blame the flooding on the levee system that squeezed the Mississippi into narrower, swifter channels.

But the huge causative element was the rain.

By July 10, the Mississippi was topping flood controls in Hannibal built to contain a 500-year flood, and 7.5 inches of rain fell in just a few hours in Iowa. The Mississippi rose to 11.2 feet above flood level, where it met the Missouri.

On July 13, all of Iowa was declared a disaster area, and manhole covers in downtown Des Moines were blowing out under pressure of runoff. Vice President Gore, in Illinois to tour the disaster, had to duck as his boat passed under power lines, the flood was so high.

Hope rose and fell. On July 16, forecasters announced that the Missouri River had finally reached its peak at 34.6 feet—12 feet above flood level—in Jefferson City.

The worst was over, said the jubilant forecasters.

A day later, rain began falling again.

On the 19th, newspapers reported that the Mississippi was dropping. After cresting at a record 46.9 feet, it was down an inch.

On July 20, the Mississippi rose again.

Day after day. Story after story. Near the town of Canton, Carnahan visited a rally point for National Guard troops assisting in sandbagging. He was there to speak to them, but first listened as an officer briefed them, explaining that if levees broke a mile away, the spot where they stood would soon be underwater.

"How far underwater?" Carnahan asked.

"Twelve feet."

Carnahan paused. "How fast?"

Missouri needed money, to rebuild bridges, houses, highways, levees. To pay for Porta Pottis and vaccinations and mosquito spraying and emergency food.

As it kept raining the requests grew bizarre, and the bizarre grew commonplace. SEMA got a call from the mayor of Hardin, in the western part of the state. She'd found a casket floated up to her home, she said. It seemed the cemetery was underwater.

Then a report came in of a second floating casket.

Reports rose of bodies in the river. Under the legalities governing disaster aid, the deteriorated corpses would have to be positively identified before families could receive money to bury them again.

This grimmest of liaison jobs for SEMA went to Bob Rogers, a low-key, chunky, genial, mustached man who had grown up in Ray County, had helped his mortician uncle run a funeral home when he was a boy, and had worked his way through college in a hospital emergency room.

By the time he reached Hardin, the flood had dislodged 800 graves.

"In many cases entire vaults popped out of the ground. Other times the vault broke or caskets broke open. The bodies were all tore up," he said.

Bob joined the search under way around Hardin.

"A lot of farmers would be picking bones out of their fields for months."

Each day, in boats or four-wheel-drive vehicles called "alligators," armies of volunteers, relatives and National Guard troops set out to find bodies, but "some were never found."

Even after the waters receded, the whole area would be covered with silt, and there would be a gut-wrenching smell of rotting vegetation.

"The searchers had sticks and shovels, and the Funeral Directors Association volunteered people to carry back the bodies . . . Pathologists flew in from the Smithsonian Institute, and did a postmortem on every body. They'd lay them on a table and start with the head. They'd know everything about that individual. Female. Several childbirths. Had tuberculosis. Died early . . . We had a registration session at the auditorium and everyone who had relatives buried in that cemetery filled out a questionnaire about their departed loved ones . . . Once a body was identified the families became eligible for assistance. But we had 470 that we didn't know who they were."

As the corpse count rose, Hardin officials became part of the chorus of supplicating voices asking SEMA and Carnahan to *Get them money*. They needed to pay for a forensic-quality fax machine. They wanted secretarial personnel. They needed a forklift or 580-case backhoe, 800 wooden pallets, surgical gowns and gloves, a high-power water hose, and a two-inch banding machine and big exhaust fans and padlocks. They wanted money for photo files for corpses and for hotel lodging for volunteers, and for a laminator. They wanted to rent a morgue facility, size approximately 44,500 square feet. They wanted postage stamps, fuel oil, four portable radios, an intercom system, land for reinterment, two one-ton diesel trucks, body bags, shoe covers, film to take a thousand pictures of corpses for identification.

The list went on. Each community had one.

The rivers had now reached a 500-year flood level, shattering all records.

Carnahan said. "You kept thinking it was going to be finished."

It kept raining.

Mel Carnahan might have had a reputation for unemotionalism, but when it came to dealing with Washington, "I was pretty hot. They were being *budgeteers*."

The governors of the stricken states had wisely pooled together, had decided to function as a team, rather than as individuals squabbling over aid. Carnahan even sponsored a "flood summit" in Missouri, in which Bill Clinton, Al Gore and members of the White House Cabinet met with the governors to discuss alleviating the suffering.

"I've never been in one room with so many protected individuals," Roy Temple said, referring to the security.

Despite all the speechmaking, federal disaster aid was not a blanket giveaway but a highly rule-bound variety of programs, which required specific criteria to be met before aid could be dispatched. And even if the criteria were met, there was simply not enough money to pay for all the damage being done by the unprecedented floods.

The state was going to have to fight for it.

In Missouri's case, the job of getting waivers in some federal laws because of the odd weather, finding loopholes in others, and pushing for supplementary legislation in Congress went to Jill Friedman, 29, new head of the Missouri State Office in Washington, D.C. She was Carnahan's lobbyist, a small, dark, energetic ex-head of Missourians for Choice, and ex-member of the Carnahan transition team.

"We kept hearing, in D.C., weather disasters are becoming a huge expense to the federal government. Hurricane Andrew had sapped the money. The attitude was, we're not a bottomless pit."

Jill was "pounding heads with Katie McGinty at the White House . . . The first holdup was getting consensus there on what should be proposed to Congress."

First there was a food issue. Jill and the lawyers found themselves going over food stamp laws in order to get officials administering the federal program to alter day-to-day procedures and get stamps into the hands of people hit by the emergency.

"If it couldn't be legal, we'd have to try to change the law."

Then there was the crop insurance problem. Federal insurance paid farmers for losses suffered to crops planted during the year a flood occurred. But in this case, some losses had involved crops planted in 1992 to be harvested in 1993.

"Some governors had lost early crops. Some wanted more flexibility."

For Missouri, one big fight involved which levees would be rebuilt. Would Washington help reconstruct only federally built levees, or local ones too? *All* of the levees were part of the same protective system lining the river.

"The White House was young, and there were folks in it who thought Mr. Clinton and Mr. Gore were elected with a very strong environmental platform, and they wanted to start implementing that environmental attitude right then," Jill said. "They'd say, 'how come we're building along the river? Let's think about moving those people away from it. Why should we be investing in these levees when they're going to blow again sometime, and anyway, we're screwing up how the river runs.'

"It was a huge fight."

The Missourians argued hotly that the middle of a disaster was not the time to make long-term federal policy.

The governor was doing a slow burn. He flew to Washington and met with Missouri congressmen. He discussed the levee issue with Al Gore personally. He gave a press conference on the steps of the Capitol, accusing Congress of dragging its feet. In one infuriating, humiliating meeting, he went to the White House with Jill and met officials of the White House Office of Budget and Management. He brought pictures of the shattered locally built levees.

"That meeting was one of the most tense and frustrating because they were turning down essential aid on the theory that bureaucratically, some of the levees were not recognized. They were treating these levees as if they did not exist . . . We *showed* them pictures of the levees that they thought nonexistent," Carnahan said.

"Finally we got through and they backed away."

And even after the White House decided it would support rebuilding different kinds of levees, there was the tricky issue of where the money would come from. The Department of Defense seemed an unlikely choice at first, but it contained the Army Corps of Engineers, who knew how to build levees.

It got the job.

The biggest fight, though, came when the states asked for alterations in the federal matching-fund requirements for disaster aid. The law required states to pay 25 cents for every 75 cents the federal government contributed.

But Carnahan wanted a 10 percent match. Jill told the White House policy advisers, "The state doesn't have twenty-five percent to put there."

No one was talking about climate change, or the greenhouse effect. No one mentioned the NOAA National Disaster Survey report, which blamed the rains on "major, global-scale anomalies which can be attributed to significant climate variations."

The problems were much more immediate, as, in any climate-caused disaster, they always are. Katie McGinty knew that President Clinton had to avoid looking as if he were throwing away money at the same time he was asking Congress to raise taxes. She was in her job to help change long-range policy, which included protecting wetlands. She knew the rivers would flood again and that new levees might eventually be topped.

But Mel Carnahan *needed money to rebuild those levees.*

Night after night, Jill walked home to her little Capitol Hill apartment, turned on the TV and watched the rain back home still falling, as if the whole damn climate had gone insane. She lay awake, unable to sleep, feeling an enormous responsibility, trying to figure out how to make a system work that just didn't seem designed to cope with the unprecedented weather. She flew to Missouri for meetings, looked down from the plane, and saw the Mississippi.

"It looked like a sea," a depressed Jill Friedman said.

On nights when the clouds parted enough over Missouri for the stars to come out, the moon shone down on thousands of other people also having trouble sleeping.

In mid-July the rivers were still rising. And at 4:00 A.M. one morning, in the city of Ste. Genevieve, road contractor and levee district president Vern Bauman shot awake.

"I had a plan to save the town. I was afraid we were going to lose it."

Ste. Genevieve was no stranger to flooding. Its 4,000 residents knew, as Bauman did, that the area had been prone to river rampages since 1785, when floods washed the place away. The village moved three miles upriver after that, and residents thought themselves safe. But floods kept coming. Recent ones had occurred in 1951, 1968 and 1973.

Still, those floods had been smaller. There had never been rains like this. Vern Bauman, in his home near the river, had fought floods off before, keeping two feet of water from getting inside his home in 1968 and losing the bottom floor to the last two crests in 1973. To Vern, a 40-foot crest was "minor," but that point had been passed on July 1.

Since then Mel Carnahan had declared the county a disaster area, and hundreds of National Guard troops had arrived to help build levees and assist as the city evacuated its lowest-lying areas.

After July 15, every day seemed to bring record-setting river crests. Predictions rose to 43 feet, then 47 feet.

Bauman, dressing for his predawn ride to National Guard headquarters, was a stocky 53-year-old whose trucks with their "Vern Bauman Contracting. Excavating, Hauling, Asphalt Paving" logos were familiar around the county.

The landscape he drove through in his pickup had been changed by the makeshift dikes meandering between river and town, around wooden homes, through backyards, across streets and along streams or drainage ditches. In the headlights, homes looked as if they had been built in cul-de-sacs formed by prehistoric mounds left by Indians, or perhaps some enormous subterranean worm had burrowed through Ste. Genevieve, detouring around buildings.

Bauman could see lone silhouettes patrolling on the top of the makeshift levees in the predawn. "River walkers" were equipped with radios in case a levee broke. Their job was to send out an alert as they scrambled out of the way.

By 6:00 A.M. the meeting with the National Guard was in full session, and Bauman's plan seemed pretty good.

The town would be broken into three areas, each jointly headed by a civilian and an officer. Several times a day the pair would prioritize the levees they watched over. Red code meant emergency. Orange designated an area needing work soon. Yellow went to solid levees.

"We set up emergency standby vehicles with sandbags on them. We met at seven A.M. every morning at City Hall, and at noon, with the Red Cross, food people, National Guard. Each noon we'd get the Corps of Engineers predictions for the river. Then we'd meet again at five and go over that night's plan."

It worked at first. The Coast Guard helped sandbag, as did tens of thousands of volunteers from all over the country. Prisoners from a nearby jail arrived to help.

"They saved the north end of town. Those guys worked their tails off, laying thousands of sandbags.

"But we were barely ahead of the river."

And the river was tricky. It was as if it had a mind, as if it were an animal probing for weaknesses. Unknown to levee walkers, water was working its way *under* a levee on the south end of town, pushing sand beneath it. On July 22, as the river reached a record-breaking 47.41 feet, levee walkers felt the structure beneath collapsing. Water began shooting out from under the levee. The weakening structure broke and water poured through a 400-foot-long gap.

Entire homes floated off their foundations. The Immaculate Conception Church was flooded to its eaves. The water rose 20 feet, but no one was hurt, and the bulk of the city remained protected.

The other good part was, with the levee broken, some pressure eased as the river level dropped a whole foot. The city took a deep breath of relief, but then heavy rains returned and the river began rising again.

By the final week in July the Mississippi had passed the 48-foot mark and Corps of Engineers officials predicted, fearfully, that it might reach 50 feet. Levees north of the Ste. Genevieve were falling like dominoes.

Vern Bauman had come up with a slogan, "We Can Do This," and it was posted all over town. His crews used concrete highway dividers to contain the base of the levees, then laid plastic sheeting in front of the dividers, then filled the area inside with bulk rock.

"We'd put the sandbags on top. We had pickup brigades. The VFW Hall was the staging area for the north end of the town. M&S Packing was the staging area for the south end of town. Valle High School was the staging area for the center. People filled 23,000 tons of screenings, fine limestone rock, in sandbags. The sandbags would be piled on wooden pallets, the pallets piled on trucks. Convoys, twenty-five to thirty pickups at a time, would bring the pallets to the levees. We had stacking groups and loading groups, 2,000 people filling sandbags at a time. It went on twenty-four hours a day."

Congressional Majority Leader Richard Gephardt flew in from Washington and toured the homemade defenses. The city lies in his district.

"He told me, 'Vernie, we have to get you protection.' "

Mel Carnahan flew in on the helicopter, filled a few sandbags himself, and watched defenses going up.

"It was kind of a miracle to behold," said Carnahan. "The blend of leadership between the mayor, that levee district official (Vern), and the Corps of Engineers. They were doing extraordinary things."

But the river kept rising.

"If the water broke through and you couldn't swim, you'd drown," Vern said.

On August 5, across the river, Illinois officials blew a hole in the side of a levee near Prairie du Rocher, to alleviate pressure on the town, and in response, on the Missouri side, water began rising.

It hit the 49-foot mark, a foot below the top of the levees.

It rose three more inches.

Then it rose to six inches from the top.

Finally on August 6, a crest raised the water to 49.67 feet. It was a sunny day, and as things turned out, the water began going down after that.

"I was never the same," Vern Bauman said.

Within days, Bill Clinton's new budget passed Congress, without the Btu tax. On August 12, he made his fourth trip of the summer to the Midwest, to sign legislation providing $6.2 billion in federal assistance to victims of flooding. But as waters went down throughout the Midwest, flood damage was estimated at $12 billion.

In the end, Washington would reduce Missouri's mandatory match of federal aid from 25 percent to 10 percent. Some state and local levees would be rebuilt at federal expense, although funds allocated would fall short of the $150 million requested. Carnahan's administration would, in a coordinated project with the federal government, institute a voluntary buyout plan to help move entire river communities out of the floodplain, to higher ground.

Ste. Genevieve would receive aid to build a huge new riverfront levee.

The 1993 flood was being called a "five-hundred-year event," meaning that, averaged over thousands of years, such a flood should occur only twice in a millennium.

But in a post-greenhouse-effect world, those who followed global climate patterns wondered uneasily, would such floods increase?

Meanwhile, that October, the president announced his long-awaited plan to combat greenhouse gasses.

In it, other than some mandatory energy efficiency upgrades on appliances, the plan would encourage voluntary efforts only to reduce consumption. It would offer profit incentives to industry and expand the EPA's "green lights program," through which Washington provided technical know-how to companies or local governments installing low-energy lighting systems. A new government–auto company collaboration would produce more energy-efficient cars. Other incentives would encourge industries to adopt alternative energies like photovoltaic cells or modern wind-driven turbines.

In one act to promote the plan, the administration flew Kansas hog farmer Burton Tribble to Washington, to publicize how farmers could help cut greenhouse gasses. Tribble's contribution to emissions came from methane gas released in his 20-foot-high pig manure tank, where the manure ripened into fertilizer. He told amused reporters that he wanted to put a lid on the tank at a cost of roughly $100,000, but, "I am having trouble finding financing. Hogs have a stigma."

Besides providing a soapbox for civic-minded hog farmers, the administration also offered incentives for improvements in landfills, home appliances and factories that came up with ways to reduce emissions of carbon dioxide, nitrous oxides, hydrofluorocarbons and methane. Dozens of utility companies sent letters to Clinton, promising to cut carbon dioxide emissions. Many state utility commissions announced that they would reward utility companies that encouraged efficient energy use.

Planting trees was part of the plan too. Nine percent of the greenhouse gas "reductions," spokespeople said, would come from planting forests, which store carbon.

"President Clinton's new plan to control global warming looks a lot more like something George Bush would dream up than like anything straight from the heart of Al Gore, the environmental conscience of the Administration," a *Times* editorial said, adding that it used "carrots more than sticks." But the editorial went on to call the plan a "reasonable response to a distant crisis."

When Sierra Club spokespeople called the numbers "squishy," Katie McGinty countered that the plan had been "put through the wringer." The numbers were "solid."

But after a while it would become clear that the rate of greenhouse gas growth would slow in the U.S., but the overall tonnage would keep rising.

And far from being soothed by the new plan, conservatives and industry groups regarded it as a step toward mandatory targets and timetables. After all, the environmentally friendly administration would still be in power in 1995, when the next big IPCC report on climate was expected, when climate convention negotiators would meet again, and environmentalists would *again* start pressuring the U.S. to more forcefully curb greenhouse gasses.

Accordingly, the fight grew more ugly. Industry and conservatives unleashed a whole new round of attacks on climate change science as the year ended. The Cato Institute, a conservative Washington think tank, published "Apocalypse Not: Economics and Environment." It called evidence of global warming "ludicrously small" and claimed that curbing greenhouse gasses would require "a degree of bureaucratic control over economic affairs previously unknown in the West."

Dixie Lee Ray, former chairman of the Atomic Energy Commission, published another book calling carbon dioxide an "unlikely candidate for causing significant worldwide temperature changes."

Conservative lobbying groups quoted Tom Karl, a scientist respected by all sides, who speculated that warmer nights being recorded around the world could have purely natural causes. Karl still awaited more evidence before committing himself one way or the other on the greenhouse effect.

But the fight was also moving away from science, to the personal. It was about to get nasty in a different kind of way. As 1994 began, strategists on both sides targeted the researchers, who were accustomed to

academic or scientific disputes but had not yet realized the direr consequences of speaking up in a political arena. They might have known
about backbiting at MIT or at Lawrence Livermore Laboratory, but
they were as stunned as lambs before a wolf pack when lofty figures in
the White House or in Congress, people *who had never even met them*,
announced to the country that they were frauds.

Now scientists would begin fearing for their reputations, jobs, and
for funding at their institutions.

In politics, if you can't disprove your opponent's position, go after
the person.

In the White House, Al Gore picked up the phone.

Washington . . . Sudan

I T WAS THE SPECTER of politics at the highest levels interfering with science that intrigued the newsman. Ted Koppel, award-winning anchorman of ABC's *Nightline,* was in his Washington, D.C., office, preparing for an upcoming show, when the vice president called in January 1994.

Outside, oddly warm January temperatures had Washingtonians, who would normally be scraping ice off their windshields, strolling out on DeSales Street wearing nothing heavier than suits and ties.

"It was the first and only time Al Gore has called me," Koppel said.

Not that he was a stranger to people proposing shows on particular subjects. The head of *Nightline* or his staff often talked to lobbyists, corporate leaders, congressmen or occasional senators who suggested topics for coverage.

"Do they have a political motive? Probably. It has never yet occurred to me that people propose stories for *Nightline* because they are enchanted by my boyish smile."

For one hour five nights a week, *Nightline* explored major issues of the day, reaching 6.8 million viewers, including policymakers all over the earth.

"You can be pretty sure that Senator X is watching *Nightline,* the president's top advisers and maybe the president," Koppel said, not bragging, just stating a fact.

His beat was the world. Global leaders eagerly accepted invitations to answer Koppel's off-the-cuff questions delivered in his politely dogged style. His grasp of issues had made the show, contrary to early expectations that any late-night news show would fail—drowned in ratings by TV movies or the *Tonight Show*—*just* as popular as the *Tonight Show*.

Now, on the phone, Al Gore told Ted Koppel he had an idea for an investigation of several scientists.

The two men had chatted briefly at times, over the years, at large Washington functions, but until now they'd only had a single lengthy conversation, years before, when then Senator Gore had run into Koppel at La Guardia Airport, as they both waited to catch a shuttle flight to the Capitol.

"We went to have a cup of coffee together, and the next thing I knew, Al Gore was sketching out for me on a napkin what was happening to the ozone layer at one or both of the poles. It was the first time I'd heard of it."

Koppel had not known what to make of the issue.

"They don't call me Doctor Science around here."

But the senator kept drawing pictures and explaining about how human-made chemicals were causing the ozone layer to weaken.

"This was clearly something he cared about. He was very earnest. It was clear he'd spent a lot of time on it. I must say I was interested if only because it did not seem like a particularly meaty political subject, but I didn't do a story on it."

Now, in January 1994, the vice president proposed a show of a more personal kind.

"He suggested that we might want to look into connections between scientists who scoff at the so-called greenhouse effect, and the coal industry . . . the Reverend Sun Myung Moon's group, and Lyndon LaRouche's organization."

The inference was that the scientists were little more than paid agents of the organizations funding them. Gore told Koppel that his office had documents proving the connections, and he offered to show them to ABC.

Koppel was intrigued, in a preliminary way, "because the vice president seemed particularly eager to point out the association between this one scientist—Fred Singer—who had been critical of the theory, and his connection to the Moonies."

Koppel told Gore the same thing he told anyone who proposed a program. "I said that's interesting, but there are two things I'm going to do. I'm going to explain to people why it is that we're doing this program. That *you* initiated this. And secondly, whatever information you have in your office . . . we're going to check it out."

Gore said that was fine. It was quite possible at this point that no show would develop. That would depend on what Koppel's staff discovered when they went to work.

As for the newsman's beliefs about global warming, he "hadn't a clue" if it was real. As for the warm January weather outside, which had many Washingtonians speculating about the greenhouse effect, "As long as I can remember, there's always been a balmy day in January. And then it gets cold again. Or there'll be three or four winters in a row where it seems we get no snowfall whatsoever in D.C., and then just when the words 'greenhouse effect' escape our lips, eighteen inches of snow get dumped on us and we have one of the coldest winters on record.

"I have a distrust of consensus," Koppel said.

The truth was, "If someone from Gore's office had called instead of Gore, I probably wouldn't have paid any attention."

Now, one of the show's producers and its three researchers read Gore's documents and phoned the scientists named in them and the organizations allegedly funding their papers, speeches, books or films. The scientists had no problem admitting to the funding.

One of them, Pat Michaels, editor of the *World Climate Review*, which received money from Western Fuels, Frederick Palmer's organization, told the show on camera, "Whether or not Western Fuels funds the review has nothing to do with what was published long before they funded it."

Koppel also watched the film *The Greening of Planet Earth*, which Palmer had helped fund too, and from which Rush Limbaugh frequently quoted. Then *Nightline* got hold of footage of a 1992 Senate hearing in which Al Gore grilled Dr. Sherwood Idso, on whose book

THE COMING STORM · 175

the video was based. Had the movie been financed by the coal industry? Gore demanded.

"It was," Idso said.

So the links were there, as Gore claimed, but did that affect the quality of the science? "What do we have here?" Koppel wondered, deciding to do the show. "Is this a case of industry supporting scientists who happen to hold sympathetic views, or scientists adapting their views to accommodate industry?"

The most interesting case involved Dr. Fred Singer, who Gore claimed received funding from the Moonie organization and consulting fees from Exxon, Shell, Arco and Sun.

Singer admitted that yes, some of his income came from those places.

But he also said, "Every environmental organization I know gets money from Exxon, Dow Chemical and so on . . . if it doesn't taint their science, it doesn't taint my science."

To Koppel, Singer seemed like the best example of what the real focus of the show should be, and it wasn't going to please Al Gore. In fact the more Koppel thought about it, the more Gore's call to him "didn't seem right to me."

And so, on February 28, 1994, as White House aides watched first in expectation and then in embarrassment, Koppel sat in his air-conditioned studio, in a blue wash of lights, glancing down at a looseleaf binder screwed into the lectern and telling his viewers, including an understandably jubilant Fred Palmer, that Fred Singer had published three books through Reverend Sun Myung Moon's Unification Church, but he had also taught environmental sciences at the University of Virginia and had been a deputy assistant administrator of the EPA during the Nixon administration, and chief scientist at the Department of Transportation.

"He had credentials," Koppel had decided.

Singer had even appeared previously on *Nightline*, during the war with Iraq, to dispute Carl Sagan's predictions that soot from Kuwaiti oil fires would rise into the upper atmosphere and be dispersed around the world.

In that instance, "Dr. Sagan was wrong and Dr. Singer was right," Koppel announced.

"You can see where this is going. If you like Dr. Singer's views on the environment, you mention his more impressive credentials. If you don't, it's Fred Singer and the Reverend Sung Myung Moon."

Koppel pointed out in the show that greenhouse skeptics comprised a small segment of the scientific community, but he also concluded with, "There is some irony in the fact that the vice president, one of the most scientifically literate men to sit in the White House in this century, should resort to political means to achieve what should be ultimately resolved on a purely scientific basis . . . The issues of global warming and ozone depletion are undeniably important. The future of mankind may depend on how this generation deals with them. But the issues have to be debated and settled on scientific grounds, not politics."

Gore's smear strategy had failed this time, but by 1994, as the new administration actively sought a greenhouse gas abatement policy, and as powerful energy corporations poured more money into fighting it, *both* sides began resorting to mudslinging. Even the scientists joined in.

Jim Hansen's entire motivation for making public statements was "to get funding for his lab," critics said.

Pat Michaels was "acting like a lawyer, not a scientist, paying attention only to facts supporting his position."

The IPCC was "controlled by a small group that had a political agenda."

Idso was "lightweight." Singer was "evil." NOAA climate labs were filled with government scientists (the word "government" always uttered with disdain) slavishly trying to curry favor with the Clinton–Gore White House, even though the *same* scientists had said the *same* things when George Bush ran the White House.

Tom Karl, though, still remained respected by all sides, and Karl had still not made up his mind.

Down in Asheville, he was concentrating on purely scientific questions, keeping fanatical track of research on greenhouse issues, both in his own work and as a lead author of a chapter in the second IPCC report, due out in 1995.

Since the first report had been published, Karl's staff had been studying weather extremes, and if they might be related to global warming. They'd been painstakingly matching real weather records to model simulations, concentrating on four key areas that Karl felt might determine whether humans were influencing climate.

Karl's staff had pored over U.S. and global records of minimum temperatures far above normal, of heavy rainfall totals between Octobers and Aprils, of severe droughts between Mays and Septembers, and they had also calculated the percentage of overall precipitation coming from sudden, short, extreme storms over the years.

The extreme events *had been increasing*, they found. The team was convinced there was a 90–95 percent chance that greenhouse warming had caused the changes, but Karl told the *New York Times* that a 95 percent possibility was not enough upon which to base a conclusion.

After all, there could be some other reason for the changes that had not been pinpointed yet.

He needed to analyze more data.

At the same time, at weather labs in England, Germany, Russia and the U.S., just as Jim Hansen had predicted, researchers found that global temperatures had begun to rise again as the Mt. Pinatubo cooling dissipated.

Nineteen ninety-four would turn out to be the fourth hottest year on record.

As the mercury climbed, drought hit Australia, New Zealand, Japan and sub-Saharan Africa.

The skies dried up. The sun grew hot.

In Sudan, the death toll began to rise.

———

Dawn came and starving Dinka tribespeople began emerging from the tree line. They'd walked from villages as far as three days off to reach this field today. They fled the Sudanese army. They said their crops were dying. They told aid workers that rebels fought each other in the east, burning huts, stealing cattle, leaving corpses for dogs.

Soon 3,000 people filled the parched field, sitting quietly on their haunches, surrounding an enormous pile—60 yards in diameter, five feet high—of 110-pound sacks with the letters "USAID" stenciled in front. Inside was Kansas-grown sorghum that had been shipped to Kenya, loaded onto trucks for transport to a UN camp near the border, and then air-dropped several days ago from cargo planes as tall, half-famished Dinkas watched the gigantic airships tilt skyward, open their back doors and send sacks tumbling out to hit the field and send up a cloud of white dust.

If direr climate projections were right, days like this would come more frequently in Africa's future—and the problems they caused would face the whole world.

The drought was worsening in rebel-held areas of southern Sudan. By 9:00 A.M., just at this particular food handout location, one of hundreds throughout the sub-Sahara that day, the crowd had grown to 5,000, with more men, women and children materializing out of the vast inverted bowl of landscape: veldt, savannah and thorn-tree-dotted horizon.

On the fourth side, in the distance, was a village of conical huts made of mud and grass. Maize grew thick in some gardens, but had not ripened yet. Cows, the prize possession of richer Dinkas, grazed, guarded by boys carrying spears.

In the field, the grim-looking men wore wraps faded to an earthy brown, rags long devoid of their original color. They carried two spears, one for fighting, one for fishing. The women, catching the eye of one of a half-dozen relief food monitors present—young men and women from the U.S., Kenya or England—beckoned with the same gesture. They pinched their bare breasts with thumb and forefinger, to signify no milk for a baby. They rubbed their palms across their bellies, to signify "no food." Naked boys were covered with gray cow dung ash, a natural insecticide. The only manufactured item in sight among the Africans was an AK-47 over one guerrilla's back. A rank hot smell thickened as the sun grew hotter, and the crowd—clustered by village, each group headed by a chief—inched closer to the food, their hunger as palpable as the buzzing of flies, the front row a border between order and anarchy.

Todd Stoltzfus, 25, of Olathe, Kansas, population 40,000, walked the perimeter of the grain bags, talking quietly to Dinka chiefs, making sure they kept their villagers orderly. He was trying to figure out how, when the distribution began, to keep the whole place from degenerating into chaos. Todd was in charge of the distribution for World Vision International and knew that in 1994, according to UN reports, 22 million Africans were at risk of starving from drought. The thousands converging on this field constituted only a minuscule fraction of them, but they also hinted at Africa's even more widespread worst-case scenarios of the future.

He was a sandy-haired, close-shaven, fit-looking man wearing a short-sleeved shirt, khaki trousers and a tractor cap against the sun. A religious Christian but not a proselytizer, he was in Sudan as part of "Operation Lifeline," a negotiated agreement among all sides in the Sudan civil war, under which 36 private relief organizations, UNICEF and the World Food Program brought supplies to civilians.

Back in Nairobi, Todd had seen warehouses piled with mountains of relief grain, boots donated by China, bicycles from India, pots from Egypt, jerricans to carry water, blankets and chloroquine and plastic roof sheeting and oral rehydration salts and benzyl benzoate.

It looked impressive in the warehouses, but he knew it wasn't enough on the ground.

After all, in this one location alone, by 10:00 A.M., the crowd had swelled to 7,000.

The pressure of dealing with it all could get to him sometimes. Just this morning, back at the World Vision camp, several hours drive along dirt roads and footpaths (there were hardly any roads), he had prayed at the daily devotional, "Lord, this devotional stems from yesterday's comment, what's the point of us being here, if every time we evacuate, everything falls apart? Look at the Bible. Every time the Lord asks anyone to do anything they say, 'Not me, Lord. I can't do it.' But they do."

The open air camp consisted of roughly a dozen green tents surrounding a dining tent, with a Sudanese village—conical huts and dried-out maize gardens—scattered outside the thorn bush perimeter, built to repel thieves and lions.

Listening to him over their Bibles had been the rest of the World Vision crew, most of whom would accompany Todd to the food distribution. There was an Ethiopian nurse whose husband had been killed during the overthrow of that country's most recent government. A U.S.-born health coordinator. A Kenyan agriculturalist, here to teach planting techniques to villagers. There were Ugandan and British-born food monitors, whose job it was to record shipments, count bags, make sure food reached its destination.

Which, today, was this field.

It should all go smoothly, Todd hoped, eyeing the crowd, *if they don't riot.*

If rival Nuer tribe militia don't attack.

If government troops don't show up.

But then one man stood, and pointed at the sky in terror.

"Antonovs!" he screamed, referring to Russian-built planes used by the government for bombing.

In seconds, eight thousand people were screaming, "Antonovs!"

The crowd launched to its feet, and fled. Mothers clutched infants. Boys tripped, running and dodging adults. The sound of 16,000 bare feet pounding on dry earth mixed with the sight of waves of brown dust rising over trampled grasses and piles of sacks.

Todd ran too. The food was left behind.

———

Sudan is the biggest country in Africa. Just the southern half is the size of France. South Sudan—the rebel zone—borders Ethiopia, Kenya, Uganda, Zaire and the Central African Republic, all of which receive its refugees. And the roots of the 30-year-old civil war were a tangle of historic ills intensified, CIA officials said, by drought.

The greenhouse effect hadn't had a thing to do with the fact that, for centuries, North and South had developed separately and that British occupiers had forbidden economic or political traffic between the more developed, Islamic North and the poorer, Christian and animist South. Or that the civil war had erupted in 1956, after independence, when the South felt ignored when it came to development, and fearful of being ruled by Islamic law.

By 1994, after numerous truces and resumptions, three factions fought each other: Dinka tribe rebels, Nuer tribe rebels and the government.

Add a drought and global warming, scholastic and military studies to be published over the next few years would say, and another key ingredient heightening anarchy was in place. "South Sudan has become one of the world's worst humanitarian nightmares," said U.S. Assistant Secretary of State Herman Cohen. "It is a tragic place where civil war, disease, homelessness and hunger form a tragedy for millions."

Just by 1994, conservative estimates of casualties topped a million, and 2 million more displaced people lived in refugee camps in the north, 600,000 in the south.

The market economy had collapsed in South Sudan. People lived by barter.

Plus, the drought was growing worse.

And now, at just this one small food distribution point, Todd Stoltz-fus realized that the panicked cries had been wrong. Government bombers had *not* materialized in the sky, and frightened Dinkas began hurrying back toward the food, fearing it might have been stolen while they were gone.

"Thieves started the panic," a rebel official told Todd. "They tried to take the food. We arrested them."

But the mood had grown even tenser because of the scare, and the anxiety rose further as groups that had arrived early in the morning watched their share go down by the minute, as more people appeared.

A 13-year-old boy from a village called Warrap told Todd, "The women were too weak to come. We'll bring food back for them. I haven't eaten in two days, and my last meal was leaves off trees."

Another boy said, "Lions eat people sometimes when they walk to feeding areas. Hyenas too, but they're more cowardly. They come at night, when you are asleep."

"Why don't you eat the cows I see," one aid worker asked a rebel official, a 40-year-old man wearing a brown tropical leisure suit, wheeling a Chinese-made bicycle and carrying a bicycle pump like a swagger stick.

Because they are the last thing to go, the official said. Because when you kill a cow you have to give so many parts to your family that you only get one meal in the end. Because cows belong to rich people. Do rich people in *your* country give away all their wealth in hard times?

The official moved off as still more Dinkas—thousands of them—kept arriving and the frustrated chiefs began arguing with Todd about distribution. The chiefs—tall, furious, armed men—demanded control of the sorghum designated for their villages. But Todd was fearful that they might not distribute it evenly; that they might use it as a tool to enhance their authority. In Somalia, where he'd been posted previously, he'd seen chiefs steal food from their own people, and he did not want that to happen here.

Instead, Todd proposed giving out food through the women. The chiefs were furious at the idea.

Then the disagreements spread to include how *much* food would go to each group. It was crucial to have a correct count of villagers, in order to distribute the food fairly. As the crowd grew more unruly, the chiefs strode through their groups, counting by hand, while Todd and a Ugandan food monitor tabulated numbers with small portable calculators.

In one place, a chief told the food monitor, "A thousand people came with me."

"I only count four hundred," the food monitor said.

"What about the women we left at home!" the chief burst out. "What about the ones too weak to come!"

There were no police to keep order. No rebel troops. Just a half dozen unarmed food monitors, the unspoken threat of aid withdrawal, and the wills of the chiefs or one or two rebel officials to keep the crowd, now about 15,000 strong, under control. The best thing would be to get distribution over with quickly, but there were intelligent rules for this, Todd knew, and an obligation to those who had not arrived yet—and who might be even now walking the last mile to reach the field.

Todd walked off to the World Vision Toyota Land Cruiser for a sip of water. Drinking water was not a good idea in front of 15,000 people who'd been sitting in the sun for the last five hours. He was tired, and ready for a change. He had applied to Yale, and if he was accepted, he told a reporter in the field, he would go.

But he would never forget what drought and war could do in Africa. He'd started for World Vision as a food monitor, paid the tax-free overseas salary of $1,100 a month. He'd been sent immediately to a town called Bor to give out food, after a faction fight.

"I'd drive past bodies in various stages of decay. I'll never forget the smell. The bodies would be in one position one day, and a different one the next. Dogs ate them."

Todd had been bombed twice at World Vision Centers. "The air pressure changed when the bombs exploded." He'd gotten malaria three times. "Everybody gets it." He'd always wanted to do relief. To work for World Vision, Todd had been required to sign a statement confirming that he was Christian, but he never proselytized. USAID rules prohibit handlers of government food aid from doing so. And anyway, he said, his faith was of a more personal kind.

And now, after a five-minute break, Todd returned to the field, where food monitors estimated the crowd had swelled to 18,000.

It was still too soon to give out food.

Back near base camp, beneath a large tree, World Vision Sudan health coordinator Dorothy Scheffel and nurse Lydia Gitonga were establishing a "wet feeding center," where children under five who qualified received a protein-rich gruel.

Two hundred mothers had shown up for a kind of malnutrition audition, carrying their babies, hoping they would be eligible for the gruel.

"There's a standard for how much a child should weigh, depending on height," Dorothy said. "We only admit children with less than eighty percent weight for height."

It was swampy hot. While the mothers sat beneath the tree, watching, Dorothy's Sudanese assistants laid a wooden board—to measure height—on the ground. They hung a scale from a low-hanging branch, and suspended a blue diaper from it. When the first baby was lain on the board, it began screaming, and the fear rippled through the other children.

The crying spread as the assistants placed the little boy in the diaper, where it hung like a suburban infant in a sleeper, except in this case, the black needle on the scale would determine if the boy ate or not.

The little boy turned out to be sufficiently malnourished so that Lydia attached a red bracelet to his wrist. He would receive the protein-rich gruel twice a day.

The baby girl who followed received a blue bracelet, meaning she would also get food, but less.

On and on they came. Children with buttocks as shriveled as a 90-year-old's. Children with lesions the size of mice on their abdomens. Children whose bellies had swollen the shape of soccer balls. Children who looked fine for two-year-olds, except they were five years old. Children whose knees were wider than their thighs.

On this one afternon in 1994, during drought, food was trickling up by convoy from Mombasa to Sudan. Food was being handed out in refugee camps in Uganda, Zaire, Ethiopia, Kenya. Food was the subject of arguments in Congress, over aid. In a million huts, in a dozen countries, millions of people were on the move to escape drought. Too many people cried out for food.

Back in the big field, it was late afternoon, and an estimated 20,000

Dinkas surrounded the USAID bags. Todd finished hammering out his distribution agreement with the angry chiefs. As they had demanded, they could hand out food. But Todd had gotten them to agree that only women could receive it.

The instant the distribution started, though, the system fell apart. Whatever force had held the crowd back disappeared as bags began to leave the pile. Restraint gone, men clubbed each other with sticks, screaming, *"Get away. This is mine."* Dust exploded around their feet as their arms rose and fell in a threshing motion.

Packs of women fought over individual bags, pulling them in different directions. When sacks ripped, children crawled amid screeching combatants, dodging blows, scooping grain into wooden bowls.

Under one tree, the group from the village of Warrap encircled their take like miners shielding a gold strike.

"You were late!" one man shouted at another. "You don't get any!"

"Everyone gets something!" another man screamed back.

It was impossible to stop it, too late to change it. After a while, when the bags were all gone, a bent old woman emerged from the melee, spotted Todd and limped toward the the young American in the baseball cap, making the gesture that had become so familiar by now: hand stretched out, circling her empty stomach. Mouth chewing air, meaning *I have no food.*

There was nothing he could do. The grain was gone. For thousands, it would mean meals for the next two weeks. For the old woman, today's distribution might never have occurred at all.

Dazed, she wandered off, and a young boy approached Todd, said nothing at first, merely stood beside him, watching the anarchy. The boy was well fed, and obviously came from a wealthier family. He glanced up at Todd, and when he finally spoke, it was in excellent English.

"The strong take from the weak. The people are fighting," he said.

No rain fell the next day. Or throughout the month after that.

If IPCC predictions are right, much less rain will fall in the years to come.

———

Six years had now passed since Jim Hansen had appeared before Tim Wirth's subcommittee, and for millions of weather watchers around the world, the notion that humans were meaningfully changing the atmosphere had taken hold.

The fact was, temperatures and extreme weather events had been breaking records with frightening regularity over the last few years.

In homes from Seattle to Shanghai to Sydney, people peered at the sky and shook their heads, and told each other, at dinners, in cars, on camels, on fishing dhows, *something is wrong with the weather.*

It was this growing dawning perception, much more even than predictions by computer models of climate change, that began to drive the political process. There was mounting anxiety that something momentous had changed. The scientific studies were there, but it was people's own instincts, their own observations, that began to affect the politics of the greenhouse effect.

Jim Hansen had *said* back in 1988 that at some point average people would start discerning a difference in climate. Now people of all walks of life were concluding that point had been reached, that the weather rhythms they had known all their lives had been disrupted, whether or not the National Science Foundation confirmed it 100 percent or not.

The more educated lay watchers, who followed research, knew, for instance, that the number of hurricanes had risen along the east coast of the United States. They were also aware that hurricanes came in cycles, as part of natural variation.

So perhaps they could tell themselves the rise in hurricanes wasn't a permanent change after all.

But the number of reported tornadoes had *also* gone up, although with Doppler radar's enhanced abilities to "see" tornadoes, some experts theorized, maybe more were simply being spotted now, so there was no actual increase.

But then there was *also* the fact that global temperatures had been breaking records, and that the current decade was the hottest on record. Of course, some experts replied, temperatures had risen and fallen periodically throughout history and a strong likelihood existed that eventually the record-breaking temperatures would subside.

Very possible.

Except extreme rains and droughts were *also* on the rise, as climate-predicting models had said they would be, in a warmer world.

Plus, numerous "hundred-year storms" and "five-hundred-year

storms" had occurred in just the last few years, *and* more ice than usual seemed to be breaking off the Antarctic ice sheet, *and* the ocean was rising, even if some experts pointed out it had risen only a very small amount.

By 1994, for millions, the bottom line was, you could always find arguments against human-induced global warming, despite the examples of extreme weather, *but why were all these things happening at the same time?*

Whether this growing anxiety represented chilling and coalescing mob power or canny human instinct, the coming storms were now about to begin affecting major corporations—some still opposing greenhouse gas mitigation politics—in a new way. Corporate attitude toward climate change—the fear of energy taxes or cuts, the abhorrence of government regulation, the doubt about the science—was a key driving factor behind opposition to political action. If big business changed its attitude toward Jim Hansen's theory, many of the impediments to action would vanish.

In late 1994, for the first time, fossil fuel companies began hearing public warnings about their future profits, because of global warming. For the first time investors were cautioned for reasons of profit against putting money into fuels that might exacerbate the greenhouse effect. Sooner or later, some financial experts predicted, fear of global warming would begin to creep into how people voted, purchased products, invested money and brought legal suits.

In November 1994, those concerns were typified in a report by Mark Mansley, a former director of Chase Manhattan Bank, for a group of financial consultants.

Mansley predicted that it was only a matter of time before public concern about the greenhouse effect grew so great that governments could be forced to limit fossil fuel use, tax it and promote alternative energy forms over coal and oil.

The report cautioned potential investors against putting money into long-term coal, gas or oil projects because they might not be as profitable 30 years down the road, given trends relating to the greenhouse effect.

Mansley also raised another threat to fossil fuel companies. If the science linking fossil fuel use to global warming became more definitive, and as real damage from global warming mounted, it was quite possible that just as angry consumers had sued the tobacco industry when scientists finally linked tobacco to cancer, an enraged public, remembering industry obfuscation during the global-warming legal process, might choose litigation as revenge.

At the same time that Mansley's assessment came out, internal memos in the reinsurance industry were raising red flags about climate change, too.

Back in Zurich, where businessmen are famed for nonalarmist tendencies, Swiss Re officials had been worrying more that the climate might be changing in ways that could permanently damage their company, industry and lives.

Andreas Schraft, a 36-year-old ex–civil engineer and member of the company's "natural perils group," had been looking into losses from storms, as insurance payouts had massively exceeded premiums collected over the last few years.

Suddenly two centuries of weather records that had helped set rates had been unable to help Swiss Re make profits. Pressure had been mounting to find out why. Was the climate actually changing, or was there some other explanation? Schraft and the Natural Perils Unit had undertaken a crucial search to isolate the cause of the losses within the company's legal department, property department and product management department. Had building codes changed? Population factors? Ways of calculating premiums?

"In the beginning we were all very skeptical if climate was even relevant to the problem," he said.

But the more he talked to meteorologists at the Swiss Institute of Meteorology, pored over articles, books and reports, the more he blamed climate. He had no particular desire to blame perils on climate. In fact, as a student at the Federal Institute of Technology, he'd studied water inundation of towns near two Swiss lakes and concluded that public perception that the lakes were rising was false. Building patterns had encouraged the flooding in that case.

But when it came to global warming, the prudent, polite, pear-shaped father of three, a respected junior executive in a restrained corporate culture, reported to top management that climate change *was* so dire a threat that he urged the company to send a representative to the upcoming climate summit in Berlin, to monitor proceedings, meet with politicians and push for action.

Management decided to send Schraft.

The company also distributed his "Policy Paper on Climate" to thousands of employees, who were now to take it into account when doing business.

It began quite graphically for a Swiss business paper: "There can be no doubt that natural climates . . . can be changed on a worldwide scale by human intervention."

It continued, "It is probable that the growing natural greenhouse effect and its impact on the climate system will lead to an atmosphere that is warmer than at any time in the past 10,000 years. The effect of global man-made climate change can only be avoided if greenhouse gas emissions are significantly reduced."

In Swiss Re offices around the world, the people who invested company money and calculated insurance rates read in the report, "We conclude that climate change, regardless of the type or cause, is not a future danger but an acute current one . . . losses resulting from climate change could lead to unmanageable proportions . . . protection against extreme weather must be reconceived . . . climate change is a methodological, political, social, technical and cultural problem which cannot be delegated to one single institution, for example research or politics. Decisions and actions have to be taken today, even if there is still some uncertainty as to how the climate will develop."

Alternatives to action, like waiting for more research, wrote the even-tempered student of natural disasters, from his sunny office near beautiful Lake Zurich, were far too risky to contemplate.

"As things now stand," Andreas informed receptive bosses, co-executives and Swiss Re staff, "we will only be certain when the feared effects have already occurred."

New Delhi ... Berlin ... Chicago

KAMAL NATH, THE FIERY and internationally respected Minister of Environment for India, had dozens of problems far more pressing than global warming on his agenda when he received Tim Wirth's urgent summons for help in April 1995.

From his office in New Delhi, a city so choked by smog that hotel clerks caution guests, "you will get sick from breathing if you take a walk," and where the airport is regularly closed because pollution clouds make takeoffs impossible, Nath oversaw the vast and inefficient environmental regulatory system in one of the most populous nations on earth.

Global warming? In India and other major developing countries, which would eventually have to cut coal use if greenhouse gasses were to be curtailed, few officials had the time, money or desire to deal with an issue they had not even caused in the first place. After all, most of the man-made CO_2 in the atmosphere came from developed countries and had been accumulating since long before India started industrializing.

Kamal Nath made that point whenever any starry-eyed U.S. environmentalist urged his needy country to cut energy use to "help the planet." India needed *more* energy, not less, and the bulk of it would come from coal, which might be dirtier, but it was also cheaper. And if the West wouldn't help make Indian coal plants cleaner, as the Rio Convention urged, India didn't have money to do it.

Despite the fact that a cleanup of Indian environmental problems would coincidentally address global warming, the country lacked resources to do it even for local reasons. Forty thousand Indians died each year from air pollution, a World Bank report said. The aged cars and three-wheel pedicabs clogging India's cities threw noxious clouds into the air so thick that they reached the Maldives, hundreds of miles away. A proposal to ban cars over 20 years old had been dropped after protests from cabbies. The country lacked a catalytic converter law to make auto emissions cleaner.

But Nath's problems went far beyond vehicular emissions. Pollution control officials faced death threats when they tried to force industries to cut emissions, according to published reports. Raw sewage and oil spills contaminated hundreds of miles of Indian coastline. And rivers were so dirty that many had bacterial counts up to 200 times above the safe limit.

Global warming? It wasn't even on the radar in Delhi.

"Former president Nehru said that the temples of modern India were dams, powerhouses, and large industries," said Dr. Ajay Mathur, of India's Tata Energy Research Institute, TERI. "Now they have been called our tombs."

Once India's needs were taken into account, though, Nath powerfully believed that "climate change will be a major issue of the twenty-first century." He had helped frame the Rio Convention, and was respected by all negotiating sides.

That was why Wirth needed his help, fast.

A trim, thickly dark-haired man, resembling the actor Joe Mantegna, Nath dressed in subdued tropical suits. He took Wirth's call in his office, which was decorated, among other things, with a large black-and-white photo of Indira Gandhi and a smaller color shot of a younger, more serious-looking Nath with Fidel Castro.

Getting on the phone, he heard the low, vibrant voice of Wirth from Berlin, where he headed the U.S. delegation at the 1995 climate summit.

"The conference is collapsing," Wirth said. "You must come."

Wirth had quit elected office. After 17 years in Congress and 5 in the Senate, he'd shocked supporters by announcing he would leave the Senate race in 1992, following a dispiriting week underlining doubts he felt about the effectiveness of the Senate, the overconcern many Senators had about perks rather than issues, and the way election campaigns are run in the U.S.

Wirth liked to tell the story of a luncheon introduction he'd given that week for Bill Reilly, head of the EPA. He complimented Reilly but felt that he "was a decent man who has let himself become compromised by losing so many policy battles inside the Administration and yet staying in office. The feeling, I realized, was not far from the view I was beginning to hold of myself."

Having quit and supported Clinton during the election, Wirth had hoped to be named Secretary of Energy or Secretary of the Interior, but instead he had become Under Secretary of State for Global Affairs.

Now he told Nath the Berlin negotiations were stalled. In 1995, if the weak Rio Climate Convention was ever going to evolve into an agreement with teeth, it was time to move to a next step. But negotiating parties were locked in the same positions they'd held in Rio three years earlier.

The U.S. wouldn't agree to targets and timetables.

The oil-producing countries opposed any agreement while small island states pushed for strong targets.

The Europeans claimed they wanted mandatory targets, but American negotiators saw hypocrisy in that position. They thought London and Bonn liked letting the U.S. take the heat for blocking an agreement that European countries would never ratify in the end, as it would be too costly.

Now, on the phone, Wirth told Nath he needed help negotiating with a whole different bloc of nations, major developing ones like China and Mexico—who feared that any agreement would be a trick to block their development and deny them energy they desperately needed.

"Tim wanted me to help find language that relieved the suspicions, so everyone would feel secure," Nath said.

He understood the distrust, though. Why should nations that hadn't even had time to enjoy major benefits from industrialization be forced to cut down on it?

◈ How could India, for instance, like many poor nations, *cut* energy when hundreds of villages didn't even have grid-powered electricity? When energy use was so primitive that "fifty percent of our energy usage comes from outside a market economy! From cutting forests!" Nath said.

In India, "headload"—the amount of wood a person could carry on their head—was an actual legal term, and 25 million tons of wood was burned by industry each year.

Perhaps *after* development had proceeded, countries like China or India could talk about global warming. Or perhaps if the developed world would *share* energy-efficient technology, help pay for changes, then it might be feasible to participate in a treaty in some meaningful way.

But as far as Kamal Nath was concerned, Tim Wirth wanted him to help convince developing countries to sign an agreement, while the U.S. refused to give anything back—not money, not technology, not a promise to cut its own emissions.

Which is what he told Wirth when he got to Berlin.

It was rainy and cold, and once again, thousands of delegates, journalists, lobbyists and NGO representatives packed a gigantic international conference center, described by one participant as looking like "a Spielberg Starship." The rainy opening days had been a kind of diplomatic Kabuki theater, "where nothing gets done and both sides play highly rehearsed parts," said Dave Garman, a member of the U.S. congressional delegation of observers, and chief aide to Senator Frank Murkowski, a skeptic on greenhouse science, who now chaired the energy subcommittee Wirth had left.

But as usual, as the conference drew toward an end, the substantive part began when important ministers—people who had real power to negotiate agreements—arrived.

Nath went into conference with Wirth, and also with the German Environmental Minister, who was afraid the conference she hosted would embarrassingly collapse.

"I asked Wirth, 'Is it your intention to stop India's development?' Of course he says no. I have great respect for his understanding of the issues. But I said, 'Tim, you've got to do something too! I must see visibly that you are putting something on the table. A timetable. *Something!*' "

Wirth, in fact, had been pushing Washington to accept timetables, but it couldn't happen now.

"Part of the environmental community wanted binding targets and we told them," Wirth said, "the end game is not going to happen in Berlin but in Kyoto (a later conference). You have to build up to it. Let's get binding targets as part of the discussion. Then we can get down to what the binding target should be."

But even promising to seriously consider targets was a problem. U.S. conservatives viewed it as a first step toward agreeing to them. "We had to honor Gore's working it through the White House, through the economics people," Wirth said.

While they waited, Wirth, Nath and Assistant Secretary of State Eileen Claussen, another negotiator, helped draw up a carefully worded "flawed draft agreement," Nath says, "which put commitments on developing countries but without any commensurate obligations on the developed countries."

It also committed the U.S. to seriously *consider* targets and timetables from now on. Because UN negotiations are a highly structured incremental process, the agreement, if accepted, would determine the nature of negotiations at the next round of discussions in 1996, in Geneva.

Distributed through the building, the draft upset Dave Garman, who spotted Wirth and Claussen in a hallway as they were about to go into a meeting where it would be accepted or rejected.

"I walked up to them and said, 'Eileen, Secretary Wirth, I feel obligated to tell you this'll never fly in the Senate.' "

At that point Wirth's cell phone rang, and while Claussen assured Garman, "We're not going to agree to this. We know it can't pass muster in the Senate," Garman had the correct impression that Al Gore was calling.

"Gore had had lunch with the president, and the president said go ahead," Wirth later recalled, but as he and Claussen went into the negotiating session, they left Garman with the notion that the agreement was dead.

"You don't tell the thugs what you're doing," Wirth said. "The guys representing the automobile manufacturers. The guys representing the GCC, they'd sit there and try to intimidate people as they walked into meetings . . . sitting there, *they* said, to let people know there was this huge coalition . . ."

The Berlin Mandate, as the new agreement was called, was a "turning point" in U.S. policy, as far as Dave Garman and U.S. conservatives were concerned. From now on, the U.S. would be creeping toward instituting energy cutbacks while big future market competitors like India, China and Mexico would be exempt. The double whammy, conservatives felt, could damage the U.S. economy while it helped competitors.

Garman, disgusted, reported home.

"They caved."

John Wilhelm, deputy commissioner for health affairs for the city of Chicago, top physician in the city, loved Saturdays. They were quiet and peaceful, and he had a terrific view of Lake Michigan from his seventy-third-floor air-conditioned apartment in the Hancock Building, in Chicago's Loop. In summers he could see sailboats on the lake, little white triangles against blue, and sunbathers on the beaches. He worked hard the rest of the week. On Saturdays he got up at 5:30, drank coffee and relaxed and read the *Tribune*.

At least up until July 15, 1995.

The nightmare started that morning while he was scanning the *Tribune*. The Cubs were seven games out of first place. Boris Yeltsin, Russia's president, was in the hospital with chest pains. Brown & Williamson Tobacco Company had known 30 years earlier that nicotine is addictive, according to the *Journal of the American Medical Association*. And of course, there was the heat.

The record-breaking heat.

Two days earlier, the papers said, Chicago had endured its hottest day on record, with temperatures reaching 106 degrees at Midway Airport. The heat index, a combined temperature and humidity reading, had hit 118 degrees. On Friday, yesterday, two three-year-olds had been killed accidentally when a day-care worker left them behind in a locked car. The city had recorded record water usage, power blackouts and thousands of hydrants opened illegally.

It was a very bad heat wave, but on the other hand, "heat waves weren't that much of an issue. Chicago has heat waves. It's always had heat waves. Our usual response worked for a hundred and twenty years."

Besides, weather reports indicated that the heat was breaking, so Wilhelm was mildly surprised when he got a call from his public relations department asking if he'd mind a reporter coming by for a brief interview.

"I asked, 'Why do they want to do an interview *now*?'"

The caller didn't know, but the cooperative Wilhelm said yes. A fastidious worker, sharp dresser and idealistic professional, he had worked for the medical relief organization Project Hope, first on its hospital ship, then in Brazil and Granada, and in Washington, D.C., as Regional Director for the Americas, before returning to his native Chicago. His proudest achievement by 1995 had been starting a West Side CDC, a local communicable disease center designed to track infectious diseases through the city.

Now he headed downstairs onto baking Chestnut Street, to meet Bob Petty, a well-known Chicago TV reporter. For the first few minutes they chatted about precautions people should take during heat waves, like drinking lots of water to keep from getting dehydrated.

Then Petty said, "I've just come from the medical examiner's. Are you aware there are seventy paddy wagons lined up outside his office, filled with bodies?"

Wilhelm, alarmed, knew no such thing. "I thought, are they from one block? Was there an explosion? A gas leak? Was it citywide? All I said was, 'I'm not aware of it.' "

Then, "I took the fast elevator to my apartment."

Wilhelm instantly called the Medical Examiner, Ed Donoghue, with whom he had attended grammar school. As he waited for Donoghue to come onto the line he realized that Lake Michigan looked different today. There were no boats out.

"And there were not a lot of people at the beach." Which was quite odd during a heat wave.

Donoghue confirmed that paddy wagons were still arriving, filled with corpses. The deaths seemed to be due to heat.

"I knew heat could kill, but all our experience wouldn't lead me to believe that seventy would die in the heat," Wilhelm recalled.

The death toll, though, had only begun.

———————

For Ed Donoghue the unprecedented emergency had begun at 10:00 P.M. the night before, when he'd learned by phone that 45 bodies awaited autopsy Saturday morning. The Cook County coroner's office was responsible for figuring out the cause of death for people dying mysteriously across an area including Chicago and 121 municipalities, home of 5.5 million people, half the population of Illinois. Normally the office handles an average of 17 cases a day, not 70. So Donoghue had arranged on Friday night for six extra pathologists to be at the office tomorrow, to handle the flood.

Then he'd gone to sleep, his practice when he knew a stressful day was coming up. "I'd learned from the crash of Flight 191 in 1979 (where 273 were killed) that you have to be well rested or you won't be able to do your job."

Donoghue was a family man with a direct manner, a quick sense of humor and a fascination with pathology—the science of figuring out why somebody died—dating back to his second year in medical school when "I was handed a syllabus and it included forensic pathology. I gotta tell ya', I didn't know what forensic pathology was."

But he knew medicine. His father had been a surgeon who took his son on medical rounds. Once Donoghue read about pathology, he was hooked. It was like working on murder mysteries.

"We're involved in 'em all the time. But the medical examiner's role is, what happened, which is different from the role of the detective, which is, who did it?"

On Saturday morning, Donoghue drove to work at 2121 W. Harrison Street through a brown hazy heat wave that had the Cook County, Illinois and U.S. flags outside the gray concrete building hanging limp as dish towels, and probably as moist. Police paddy wagons were still arriving, joining a growing line outside.

"Things were worse than I thought."

Because he had worked disasters before, he began arranging logistics. He called two extra pathologists, ordering up the full complement of medical personnel available to him. He told his administrative assistant to order extra autopsy supplies and also lunches for later, to keep morale up for workers, to keep them from driving off to restaurants when it was time to eat.

He told the pathologists, "We'll autopsy everyone under forty-five, but if they're over forty-five, and nothing looks suspicious, if the death circumstances indicate, say, windows closed, body heat excessive, hot, no air conditioning, we'll do the death certificate without an autopsy."

More paddy wagons were joining the line outside.

Donoghue knew how heat can kill. He knew that when human body temperatures reach 105 degrees, the hypothalamus, a gland in the brain, orders the heart to increase blood flow. Blood carries heat to the skin, which cools when sweat evaporates. But if a person can't sweat, or if liquids aren't replaced, the body gets hotter. Victims suffer rashes, headaches, exhaustion and disorientation. If sweating decreases further, they can fall unconscious, enter a coma and experience brain damage or die.

"When the temperature gets that high, it can cause electrolyte abnormalities, can lead to cardiac arrhythmia. The heart stops beating in its normal fashion . . . You also get bleeding problems because of numerous hemorrhages. And large lesions in the skin because the blood coagulation goes out of whack. That's another indication a death is from heat. The blood doesn't coagulate, doesn't clot."

But Donoghue didn't usually *see* many heat deaths. He'd blamed only eight deaths on heat in 1994, three in 1993, and one in 1992. Now bodies were coming in from all over the city.

Downstairs, plywood county burial boxes were turned upside down to be used as platforms for the dead. Pathologists examined extra bodies in hallways.

By late morning, police were instructed to take some corpses directly to funeral homes and Donoghue arranged for refrigerated trucks to handle the overflow. By noon volunteers began arriving from mortician schools as TV news informed the city of the backup at the morgue. Donoghue worked most of the day on bodies, and by late afternoon he met with reporters, but found their questions "third rate, boiling down to, like kids, saying, why, Daddy? *Why?*"

Body after body. Like a war zone. Mayor Daley began questioning if so many deaths could really be weather related. As Donoghue explained to Wilhelm or reporters over the phone that yes, all these deaths *were* from heat, ironically he sat in his office near one of his favorite decorations, a graph showing a study of death from extreme weather—mortality among Napoleon's troops during their frigid winter retreat from Moscow.

Donoghue felt overwhelmed. "But I was confident I'd done everything I could do. Luckily I didn't see the end of things. On Saturday we saw eighty-five, so we thought, '*we can do eighty-five*. On Sunday, *we can do a hundred*.' "

One grim bit of ignorance made things easier for him.

"We didn't know we'd go to seven hundred."

Neither John Wilhelm, Ed Donoghue or paramedic Michelle McInnis followed climate science, such as the work of Dian Gaffen and Rebecca Ross, NOAA researchers who would find within three years that "in the past half century, mean summertime temperature in the U.S. has increased, with nights warming more than days." Or that statistics and weather data from dozens of cities show that there are critical threshold temperatures above which human deaths increase sharply, or, most importantly for Chicago in 1995, "the *combined* effects of temperature and humidity in sultry weather are . . . a widely used measure of heat stress."

They did not know that between July 13 and 15, 70 daily maximum temperature records had been broken across the U.S.

What Michelle did know, driving her blue Pontiac Sunbird into work at Chicago Engine Company 122 Saturday morning, was that the road was buckling and forecasters said there was a lot of extra humidity in the air. Recent heavy rains had saturated the ground throughout the Midwest "and a high pressure area had pulled the moisture up from the ground, like a glove lying on wet grass" said Paul Dailey of the National Weather Service.

With temperatures staying high at night, there had been no opportunity, however brief, to cool off.

Arriving at the brick firehouse, she saw that Ambulance 24 was gone, which meant the previous shift was still on a run. Usually she started work by checking equipment; monitors, oxygen tanks, medicines, oil and gas supply, but there was no time for all that today. At nine the ambulance roared up and the driver said, leaping out, "It's bad out there. You have a run."

Michelle was a respected paramedic, "very good, very caring, very empathetic," said her supervisor, Kevin Sullivan. "She's soft-spoken and works hard. She's who you want if you're having an emergency in your home."

And Chicago's South Side, the area in which she worked, was poorer, densely populated and filled with older small homes and brick tenements, as well as a few middle-class neighborhoods. Her first patient turned out to be a woman who had collapsed at a funeral, from heat.

"While we took her to the hospital, the dispatcher told us, 'when you're done, you have another run.' Dispatchers were begging over the radio, 'Can anybody take a run?' "

Michelle told partner Greg Stinnet, "This is crazy."

The next run was to a house where "a man was literally cooking. All the windows were closed. He was in front of the TV, in pajamas, unconscious. His skin was burning. I asked the son, who'd found the father, 'Why isn't the window open?'"

"He was afraid of being robbed," the son replied.

Back in the ambulance, crisscrossing the flat Midwest streets, passing tenements, churches and storefronts, Michelle could hear other ambulance crews on the radio, in a frantic chorus of, "We have a DOA! We have a DOA!" so many dead-on-arrivals that the supervisors told crews to leave bodies for the police instead of tying up much needed ambulances.

It was necessary but it seemed wrong. "In one family a woman started shouting at us, 'You're just going to leave Mom there?' We had a lot of explaining to do.

"I had five DOAs that day."

The isolation of the elderly and the absence of air conditioners contributed to the death toll, Michelle saw. Older people often didn't even realize how hot they were. Their bodily temperature systems did not function efficiently. People with high blood pressure were also at greater risk, as were alcoholics and people suffering from respiratory disease, kidney disease, heart disease. Heavier people were more likely to have problems, and anyone exercising outdoors.

Michelle passed the looming tower of Mount Sinai Hospital, which was so clogged with heat cases it had gone on "bypass," like many hospitals, turning away ambulances while walk-in patients kept coming in.

"At one point we had so many ambulances here they complained we were holding 'em up, but we didn't have carts for patients," said Dr. Leslie Zun, Chairman of Emergency Medicine. "We ran out of ventilators. We were running out of disposable procedure trays. Our AC didn't have the capacity to deal with the heat, and we used fans to try to cool rooms. It got as hot inside as out. I remember looking around the emergency room, filled with patients. Some with sheets on. Some had taken them off. Some unconscious. Some stuperous. I remember thinking, *This is real bad.*

"We had plans for disasters where multiple victims come in at the *same* time, but not *over* time."

Michelle drove through deserted streets and past stalled steaming cars and by A. A. Rayner & Sons, one of the biggest funeral parlors on the South Side, where funerals were proceeding doubletime, and the three small viewing rooms for caskets on the first floor, which normally accommodated one grieving family at a time, now held two, or even three, and where the Universal Casket Company, which normally made two deliveries a week, would soon be coming every day.

Charles Childs III, A. A. Rayner's grandson and future head of the business, went to homes himself to pick up bodies. He was a polite, quiet man, an outdoorsman, a hunter. "In one house the woman had died trying to reach the window . . . We saw skin slippage, people starting to decompose faster than normally. We tried to remove clothes and skin came off . . . Stomachs swelled. Eyes swelled. Tongues protruded.

"I've never experienced anything like that heat."

By two o'clock Michelle was exhausted. She'd had no break, and the calls kept coming. The news over the radio grew more bizarre. A boatload of kids, day campers, had landed by the Shedd Aquarium, and the kids were so heat sick that the fire department was hosing them down with spray and taking some of them to the hospital.

Michelle told Gregg, "That's stupid. What's a boatload of kids doing out in the heat?"

The ambulance began having mechanical and air-conditioning problems. Twenty-eight of Chicago's 59 ambulances would be disabled at some point in the heat wave.

Michelle called her mother to warn her to stay inside. She told her two daughters, who wanted to go to a local swimming pool, to stay by the air conditioning.

"I was overwhelmed by the heat. The work. The numerous dead bodies. The ignorance of people not understanding that a blowing fan in the middle of a room does not make cool air."

But the deaths wouldn't stop in Chicago for three more days, as more bodies were discovered. The ambulance runs wouldn't stop for Michelle until the sun came up on Sunday morning, when her shift was over, and she told the incoming crew, as her predecessors had said to her, "It's bad. You have a run."

She would never hear about another heat wave starting about the same time, halfway around the world, in India. Even in the hottest countries, heat could kill, and temperatures in India were topping 119 degrees. In Delhi, not far from Kamal Nath's office, people were dragging mattresses onto their rooftops and pouring water over themselves to try to keep cool at night. They were clogging hospitals as they collapsed outside buildings, in seminars, at home, in cars.

In poor villages, across northern India, fights broke out at water pumps. Corpses lay alongside roads. "Many, many hundreds of deaths, all from heat, and not all the cases were reported," said Dr. S. K. Minocha at Doctor Ram Mohar Lohea Hospital, not far from Kamal Nath's office.

None of this was reported in Chicago. The two heat waves were not considered connected. They were "flukes." They were local.

"I went home," Michelle said in Chicago, "and I cried."

———

By 1995, Republicans had taken control of Congress, and the purpose of Capitol Hill hearings on global warming had reversed. If during the Democrat years, Senators Gore, Wirth and like-minded congressmen had used hearings to warn the public of climate dangers, Republicans now headed the committees, and hearings were designed to spread doubt over whether any danger existed at all.

On November 16, Republican congressman Dana Rohrabacher of Orange County, California, a former speechwriter for Ronald Reagan, called to order the House Subcommittee on Energy and Environment, of which he was the new head.

Rohrabacher had titled today's hearing "Scientific Integrity and the Public Trust" and said it would "look at how agencies under this subcommittee's jurisdiction are using science to formulate public policy." The hearing would also examine "the use of computer models to forecast global warming over the next 100 years."

Personally, he held the same opinion of global warming that he'd had of global-cooling theories, in the 1960s.

"I never believed any of this gobbledygook."

He thought most people promoting warming were in it for the money.

"They're getting government grants. Or selling books or going on the lecture circuit. If you're on that side of the issue, the horn of cornucopia is flowing . . . I'm not saying people don't believe this. I'm saying you can talk yourself into anything if you're going to make a profit."

The avid surfer and fan of Oliver North, whose signed photo adorned his office, Rohrabacher was a man whose occasional clownishness could be mistaken for lack of intelligence by people who later regretted it if they took him on in public.

"I made a mistake," the ex-journalist liked to say of global warming, a glint in his eye. "I took Geology One in college, to get science out of the way. In the course I had to memorize dates of the ice age, and it stuck in my mind that for *millions* of years the ice would go," he would say, pushing his hands forward as if they were glaciers, "one way, and then it would go (hands back) another way. *Long-term trends.* Slight warming. Slight cooling. They push glaciers this way and that . . .

"My mind-set is against apocalyptic prediction until it's proven right."

Rohrabacher also liked to talk about cranberries to show how he felt about global warming.

"In my life I've seen so many of these predictions not proven. Like cranberries! When I was a kid in the early fifties, there was a *scare about cranberries!* For two Thanksgivings we didn't have cranberries. I asked my mother why, and she said," he would say, brows up, playing mom, whispering in horror, "they cause cancer!

"It destroyed the income of so many cranberry farmers. All baloney. And then it was cyclamates that caused cancer. The U.S. soft drink industry that had spent hundreds of millions of dollars developing it and putting it in their product were forced to take it out. And guess what? The government finally reported they were wrong. As if that billion dollars didn't count."

Rohrabacher would screw up his face like a cute, embarrassed actor in a sitcom who has just made a mistake, and he'd say, in a nasal theatrical voice, "Ooops. Sorry!"

From Rohrabacher's standpoint, Chicken Littles like Al Gore weren't going to use bad science to turn the mighty U.S. economy upside down, and then later, when the nation was reduced to shambles, confirm a report *disproving* global warming, and say sheepishly, like those cranberry-scare bureaucrats, "Ooops! Sorry!"

Now he told his subcommittee that his goal was to "have an honest debate on the issues," but with Congress growing more bitterly divided and rancorous, the definition of "honest debate" seemed to differ depending on who you asked. By November 1995, with the Berlin Mandate in effect and the next IPCC report due out any moment, the fight was breaking into dozens of individual skirmishes over research programs, science, and over the hearts and minds of average Americans, any of whom would come away confused by today's hearing.

Pat Michaels, lead-off witness, testified that the scientific review process of the IPCC was "highly flawed."

William A. Nierenberg, director emeritus of the Scripps Institution of Oceanography, said there was a lack of certainty that *any* climate change would be harmful, and that he believed that "one can now safely wait until the climate changes become clearer and more definitively negative before taking action."

Thomas Gale Moore, senior fellow at the Hoover Institute, an economist, talked glowingly of warm periods between the ice ages, and of great leaps in human development that had occurred during those years.

"If you get the French travel guide and go around Europe and look at their three-star cathedrals, almost all of them were built during this time. It was an amazing building boom," he said.

Moore admitted that global warming would not benefit everybody. The Maldives, for instance, "are going to be in trouble," he said. "Fortunately, there are not many people in the Maldives, and we can do something about that."

Presented with an EPA report predicting rising ocean levels, Rohrabacher retorted, "I am tempted to ask what this will do to the shape of waves and ridability of the surf."

Democrats on the subcommittee seemed more kindly disposed toward believers, like Dr. Robert Watson, Associate Director of the White House Office of Science and Technology, who testified as a co-chair of the IPCC that there was "no doubt" that human activities had significantly increased atmospheric concentrations of greenhouse gasses. "We expect the incidence of extreme high temperatures, floods and droughts to increase in some regions."

Republicans liked the skeptics.

Listening to the arguments, environmentalists present felt there was more than a little irony in the fact that the IPCC had been established during George Bush's administration, with its approval, as a way of determining whether man-made greenhouse gasses were a serious issue. But now that many conservatives did not like what the IPCC was learning, they wanted to distance themselves from it.

Rohrabacher's hearing ended with the sense that little had changed in terms of the science in the last seven years, but two weeks later, flying into Madrid for a meeting where the second IPCC report would be officially approved before release, Tom Karl knew that *his* own position was changing, that *lots* more was now known about the greenhouse effect since that first raucous meeting at Dormy House in 1988.

For instance, Karl remembered that in 1990, when the first IPCC report was released, he'd agreed with it that global temperatures were rising but questioned along with it whether the change was major, was caused by human behavior and even whether model predictions were accurate as to what future changes would include.

But by 1995, Karl had been following improvements made to models in the U.S., England and Germany. His earlier worry that models did not include the cooling effect of aerosol particles like sulfur dioxide in their projections no longer applied.

"Modelers added aerosols and were able to reproduce historical records better. It gave us more confidence in the models."

With the cooling effects added, projections of future warming had *dropped* since 1990, but the upper-range projection, a 6- to 7-degree Fahrenheit rise with doubled CO_2, was still quite serious, Karl felt, and he was more inclined to believe projections now.

"Plus we knew that as countries like India and China develop they might curb their sulfur emissions because they cause effects like acid rain, that become unbearable locally."

And once that happened, the sulfate aerosols would not be offsetting warming greenhouse gasses anymore.

But it wasn't one particular study or projection making Karl less skeptical. It was a whole range of work.

For example, in April, David J. Thomson, a researcher at AT&T Bell Labs, had published a statistical climate analysis of twentieth-century records that showed a decreasing temperature difference between winters and summers in the Northern Hemisphere, corresponding to recorded global warming.

The report had concluded, "The effects of increasing greenhouse gasses may be worse than previously thought . . . simple solar variation is not a statistically viable explanation for the increase in average temperature . . . the rate of change in the *timing* of the seasons in the Northern Hemisphere is without precedent in the instrument records."

Then, in June, a team from Germany's Max Planck Institute for Meteorology had announced that there was only a 1-in-40 chance that natural variability explained the warming of the last three decades.

Then there was Karl's own work in Asheville, which showed the pattern of weather extremes rising.

But the big watershed study, which Karl had first learned of in a phone call from its author—a paper that had electrified climate scientists—had the uninflammatory title of "A Search for Human Influences on the Thermal Structure of the Atmosphere." It was a "pattern study" trying to match up trends between *real* climate throughout the atmosphere and *re-created* computer model projections after the models were adjusted to include greenhouse gas effects, cooling industrial emissions and depletion of stratospheric ozone.

The study concluded it was "likely" that the real temperature rise had been partially due to human activities.

The key author of that study, who was now also flying into Madrid as a convening lead author of soon-to-be-notorious Chapter 8 of the new 1995 report, was a small, quiet PhD named Ben Santer, who worked for Lawrence Livermore National Laboratory near San Francisco. He was new to the IPCC process and had no idea what a bombshell he was about to set off.

Soon he would be the target of a national campaign to discredit him. His excitement over the science would turn to horror over the politics. "I'd think the roof was caving in . . . I'm amazed the work I do can inspire such hatred."

On the surface Santer seemed a softie—with his carefully articulated speech, boyish cropped brown hair, neat understated dress, youthful face and attitude at times of trusting naiveté. But he was a marathon runner and a mountain climber—the last one to scale Mt. St. Helens before it blew up—who had once fallen into a crevasse in the French Alps and slowly worked his way to the surface.

His doctorate had involved evaluating climate models, work he continued at the Max Planck Institute in Germany, and at Lawrence Livermore Laboratory when he arrived there in 1992, under a program started when George Bush was in the White House.

"The Energy Department recognized that there might be changes in U.S. energy policy, so it behooved us not to rely on British or German climate models, but to have our own capabilities."

Accordingly, Santer's team analyzed the work that people like Jim Hansen were doing—analyzed the analyzers of global warming, so to speak, assessing whether the cloudy crystal balls called climate models could really do what their makers claimed.

In 1994, Santer had been asked to join the IPCC process as a convening lead author of the 1995 report.

"It'll be good for your career," the caller told him.

"It didn't turn out that way," Santer would recall.

As for the politics, "I was apolitical. I wouldn't have agreed to do the IPCC chapter if I felt I wasn't free to say what I really believed."

But the process excited him.

"It was like the quest for the Holy Grail. You're asked to do the impossible, to be completely objective, to put aside all the filters through which you view the world. You're asked to forget about all your prejudices. To just look at the science and come up with some dispassionate assessment. It's tremendously difficult. All human beings have filters through which they view the world. But I think we bent over backwards to be inclusive and look at a whole spectrum of work."

Having said yes and entered late into the process, a fact that would come back to haunt him, he and three other lead authors held a meeting at Livermore in September 1994, in which roughly 20 experts "working on the cutting edge of detecting climate change brainstormed and tried to flesh out the structure for the chapter.

"We came up with three main areas to look at. First, we had a better idea than we'd had a few years earlier what kind of climate change signal we should look for against the background of natural climate variability. Climate is like a symphony, and there are all kinds of things contributing to the mix . . . The tough job is disentangling the individual notes and figuring out how large is the human influence.

"Second, we had better estimates of *natural* climate variability. For the past couple of years the paleoclimate community had been looking at natural indicators of climate change: tree rings, ice cores, stuff like that. And they'd used this information to try to reconstruct how the climate might have changed over the last thousand years, and *that* suggested that the recent changes over the last century were unusual. Also, climate model experiments had shown that you could run climate models without *any* changes in greenhouse gasses or aerosols in order to understand the background natural variability, and these would enable us to get a better understanding of what the climate would have done *had there been no humans on the planet.*"

Finally, Santer examined reports on *patterns of change.* "Look, imagine you're sick and you have a temperature of 103 degrees. Well, that's outside the range of normal human body temperature. But it doesn't necessarily give you an idea of what's wrong with you. In order to learn that, you'd have to undergo complex diagnostics. In the climate context, people had studied changes in global mean temperature. We suspected that it had risen by half a degree Celsius over the twentieth century, but just like in the medical analogy, that number didn't tell us the possible causes of the increase. To understand that, it was better to look at *patterns* of change. The underlying assumption was that the different contributing factors—changes in the sun, volcanoes, greenhouse gasses—have different characteristic signatures, different fingerprints.

"By looking at *patterns* of change rather than one number, it became easier to disentangle the climate record."

The chapter took up more of his time as he read the literature, did his own work and conferred with other authors.

"It became obvious early on that some people had very strong ideas about detection and felt the issue had already been settled. Other voices were far more cautious."

In the draft chapter he was bringing to Madrid, "[w]e tried to summarize the three areas of progress in roughly twenty pages. The chapter cited over 135 scientific papers."

But Madrid was more than just a gathering of scientists trying to summarize their collective findings, which would have been difficult enough. Madrid was also a political meeting including diplomats from participating countries, who would also vote on the wording of the upcoming report.

Because an acceptance of Chapter 8, Santer's, would be an admission that for the first time the IPCC acknowledged human influence on climate, the fights started as soon as he and the other authors presented it.

Instantly, his sentence, "Emerging evidence points to a detectable human influence on climate," came under heavy attack from Saudi Arabian and Kuwaiti diplomats, who wanted to change "emerging evidence" to "preliminary evidence," and then, "preliminary evidence subject to large uncertainty."

A headline, "Significant New Findings Since IPCC 1990," was struck from the report.

A Kenyan delegate argued that Chapter 8 should be dropped entirely.

"The real shocker was the nuances of language," Santer said. "I insisted that there was a clear scientific basis for Chapter Eight. We'd learned a lot since the first IPCC report and we'd carefully laid out what we *did* know and what we *didn't* know."

The fights grew so bad that Sir John Houghton, chairman of the meeting, formed a special group of 20 delegates from different countries to work out differences. Then, after the full working group voted to accept the chapter, Santer was sent back to incorporate the changes into its final version.

The meeting also approved a "Summary for Policymakers" to accompany the finished IPCC report.

And that report, of which Tom Karl approved and which represented the conclusion gleaned from hundreds of papers, read, in the end, "Greenhouse gas concentrations have continued to increase."

It stated, "If carbon dioxide emissions were maintained at . . . current levels, they would lead to a nearly constant rate of increase in atmospheric concentrations for at least two centuries . . . approaching twice the pre-industrial concentration . . . by the end of the 21st century."

It said, "A range of carbon cycle models indicates that stabilization of atmospheric CO_2 concentrations at 450, 650 or 1000 parts per million could be achieved *only* if global anthropogenic CO_2 emissions drop to 1990 levels by, respectively, approximately 40, 140 or 240 years from now, and drop substantially below 1990 levels subsequently."

But CO_2 levels were still rising in the atmosphere, not falling.

The report also said that a best estimate of temperature rise in the future, if cooling aerosols were taken into effect, was two degrees Celsius by 2100, with low and high estimates being one degree and 3.5 degrees, respectively. But "in all cases the average rate of warming would probably be greater than any seen in the last 10,000 years," although annual and decadal changes would include considerable natural variability.

Missouri flood officials who saw the summary might have stopped at the line, "The 1990 to mid-1995 persistent warm phase of the El Niño–Southern Oscillation (which causes droughts and floods in many areas) was unusual in the context of the last 120 years."

Those who'd been unable to cool down at night in Chicago that July might have been interested in the statement, "Night-time temperatures over land have generally increased more than daytime temperatures."

Residents of the Maldives would have paid particular attention to, "Global sea level has risen between 10–25 cm over the past 100 years and much of the rise may be related to the increase in global mean temperature."

Officials of the World Skiing Championships, canceled in 1995 for the first time in 64 years due to lack of snow, would have lingered over, "A general warming is expected to lead to an increase in the occurrence of extremely hot days and a decrease in the occurrence of extremely cold days."

And firm believers in greenhouse theory were chilled by the conclusion, "Future unexpected, large and rapid climate system changes (as have occurred in the past) are, by their nature, extremely difficult to predict. This implies that future climate changes may also involve 'surprises,' " like more severe changes than the report projected.

Within the context of "many uncertainties" that still remained, including cloud feedback role, ocean role and questions about model accuracy, the report concluded, and Tom Karl now believed, *"the balance of evidence suggests a discernible human influence on global climate."*

Back in Asheville, he continued his studies of temperature records and was already starting work that would impact on the next IPCC report, due out in 2000. Although he could still understand the arguments of greenhouse skeptics, he was growing less inclined to believe them.

As 1995 ended, "There was still ten percent doubt," Karl said. "Maybe the correlation between models and climate was a coincidence, or maybe we had convinced ourselves there was a relationship."

And even if there *was* a definitive human–climate relationship, Karl added, "maybe the warming would not be so great."

"But if you had to put your money on a position, you would put it on that humans were affecting global climate. You don't put your money on that they're not."

Washington ... Zurich ... London

BEN SANTER'S SCIENTIFIC REVELATIONS in Madrid—called "the most important . . . in a decade" by the *New York Times*—were about to fuel a whole new round of attacks on big coal and oil, whose representatives needed to discredit the researcher and the whole IPCC process, fast. The surprise attack began as Santer finished his lecture in the House Rayburn Building—on science, not politics. He would never forget this turning point in his life.

It was May 1996 and Santer's talk was part of a series sponsored by the U.S. Global Change Research Program. The room was filled with scribbling journalists, congressional staffers and environmentalists or industry lobbyists eager to hear the man whose work had stirred up such controversy.

For the last half hour Santer had been explaining his "pattern studies," using an overhead projector to show how the atmosphere responded differently to various influences—one way to solar energy variation and another to greenhouse gasses, for instance. By isolating the patterns, scientists might quantify human influence on climate, he said.

Wrapping up the talk, he "didn't have an inkling of what was about to happen."

Soon his health would deteriorate. His marriage would fail. He'd fear for his job, friends' jobs and funding for his laboratory. But at the moment he merely asked the audience if anyone had questions.

Hands went up.

The first query came from Fred Singer, the scientist Al Gore had asked Ted Koppel to investigate in 1994. Singer acknowledged that there might be some tiny human effect on climate, but implied it would never be consequential. Santer had heard the argument before.

Then Bill O'Keefe of the American Petroleum Institute brought up the fact that a prominent MIT professor disagreed with Santer's findings, and Santer replied that "a number of respected scientists around the world were involved in the report." That sort of exchange was standard too.

But suddenly a stranger—Don Pearlman, the lawyer/lobbyist for coal and energy companies—was standing up some rows back, demanding to know who had made insidious "changes" to Santer's chapter in the IPCC report—to eliminate scientific doubt in it about global warming—between the time the chapter had been accepted by the Madrid meeting and when it had appeared in final form.

Pearlman sounded outraged. Hadn't a *whole* concluding summary been deleted? Hadn't the words "no study to date has positively attributed all or part of the change to (human) causes" been cut from the chapter?

Santer was stunned. *He* had made the cuts, but only because the same views were expressed elsewhere in the chapter, or because he'd been asked to merge two summaries into one and put the presentation in the same format as other chapters.

But now Pearlman was referring to the deletions as if they'd contained key information that Santer had suppressed.

"Who made the changes?" Pearlman thundered.

Santer, off balance, answered literally, "I did."

"Twenty percent of the chapter was *devoted* to uncertainties," Santer would say later. "But Pearlman was treating a *draft* chapter like a legal document. Like nothing should have been changed."

One journalist present, author Ross Gelbspan, recalled, "It was one of the most brutal, inhumane things I've ever seen. Santer was standing up with his slides and his pointer, showing graphs and maps of the atmosphere. They simply stood up in this large room and began making accusations. When he tried to respond, they shouted him down, and wouldn't let him complete his explanation."

Transcripts of the Madrid meeting would confirm later that Santer's chapter had been accepted with the "clear understanding" that he would modify it to "improve its presentation," IPCC chairman Bert Bolin wrote.

But somehow in the Russell Building, with journalists *writing down what Pearlman was saying,* innocent acts sounded like crooked ones, and Santer felt as if the lecture room had turned into a courtroom. He was being examined in a way he never had been before. The way the lawyer kept hammering at him implied some political interpretation of his editing, an accusation that he'd violated IPCC rules, ignored science, eliminated caveats in research in an underhanded trick to make the IPCC report more frightening.

"Who is responsible for making these changes?"

"Uh . . . I am."

"I went out of the room to cool down. I came back in and Pearlman started shouting at me again! I was getting so mad I was either going to walk out again or hit him."

Don Pearlman, asked about the incident later, remarked that he "might have asked a question, or might not have."

Santer, reeling, was as accustomed as any scientist to having his theories questioned, but not his motives.

"A lot of people worked very hard on this. I represented scientists around the world who had devoted a significant amount of time and energy toward producing the best possible assessment, and then somebody who wasn't even there says it's all a load of crap. I fought hard to *keep* the discussion of uncertainty in Chapter Eight against scientists who suggested it should be in other chapters!"

Santer flew home distraught, having no idea the attack was only beginning. Back in Washington the GCC was distributing a ten-page memorandum to reporters covering climate change, detailing charges.

"Extraordinary" changes made to Chapter 8, the memo crossing desks at the *New York Times* and major networks said, had "changed the fundamental character of the chapter." Phrases like the deleted "None of the studies cited . . . has shown clear evidence that we can attribute [global warming] to . . . increases in greenhouse gasses," had been "thought to be sufficiently important by the scientists who wrote the orginal chapter eight that they were included in their concluding summary."

The memo did not point out that Santer had written the original summary. In other words, *he had been editing himself.*

It also failed to mention doubts that Santer had included in the revised chapter and summary, like, "Our ability to quantify human influence on climate is currently limited" and "important uncertainties remain."

In short, as Ted Koppel had pointed out when Fred Singer was attacked by Al Gore, anyone following the dispute could see where it was going. If you liked the IPCC report, you could find scientific doubts in it. If you didn't like it, you picked out individual phrases that had been deleted.

Santer had been blindsided. And now more letters from Don Pearlman and the GCC began reaching journalists and congresspeople, accusing some unnamed person of "scientific cleansing," a play on the term "ethnic cleansing" then being used to describe mass murder in Bosnia. Things can't get worse, Santer figured, but they did. After arriving home he was sitting in a jury room in Oakland one day, waiting to be impaneled, when he checked his voice mail and, horrified, heard a message from a journalist calling from Geneva to warn him that the *Wall Street Journal* had just published an op-ed piece by a past president of the National Academy of Sciences, accusing whoever had changed the chapter of the worst abuses of the peer review system he'd ever seen.

"I went up to the clerk. I . . . I said I have to go. I said I can't serve on any jury. I have some very serious problems. They excused me. They saw I was pretty upset."

The op-ed piece, by Frederick Seitz, respected president emeritus of Rockefeller University, said that, "Nothing in the IPCC rules permits anyone to change a scientific report after it has been accepted by the panel of scientific contributors and the full IPCC" and "If the IPCC is incapable of following its most basic procedures, it would be best to abandon the entire IPCC process . . . Few of the changes were merely cosmetic: nearly all worked to remove hints of the skepticism with which many scientists regard claims that human activities are having a major impact on climate . . . and on global warming."

"I thought the roof was falling in," Santer said. "I felt sick. Here I was, a lowly scientist, and the author of the article was past president of the National Academy of Sciences, the most august scientific body in the U.S."

As if the op-ed piece wasn't bad enough, Santer learned that Congressman Dana Rohrabacher had sent energy secretary Hazel O'Leary a letter asking about Santer's "changes" and about funding to Lawrence Livermore National Laboratory.

"It was horrible. Now I was concerned about funding for my colleagues, my friends, their families . . . we're fully funded by the Department of Energy. If someone's coming around saying, 'Where's Santer getting his money,' it's scary."

Santer began losing sleep. He was not eating well or feeling well in general. He spent all his time trying to figure out how to defend himself. Responding to the charges was taking him away from his family.

Slowly and agonizingly, he began to fight back. The *Wall Street Journal* published Santer's Letter to the Editor in reply, co-signed by 40 scientists, as support coalesced around him. Bert Bolin confirmed that Santer had been asked to make the changes in Madrid. The Executive Committee of the American Meteorological Society published a statement saying that the attacks on Santer ". . . have no place in the scientific debate," and that the *Wall Street Journal* piece was "especially disturbing because it steps over the boundary from disagreeing with science to attacking the honesty and integrity of a particular scientist."

All the lead authors of original Chapter 8 announced that the final version was the "best possible evaluation of the evolving scientific evidence."

In a vote of confidence, the IPCC asked Santer to become a lead author for their next report, due out in 2000, but he had had enough, and he declined.

"I wanted an independent look at detection and attribution by people who were not involved as lead authors in 1995. That way, if they came up with the same or stronger conclusions, no one could accuse me of being involved."

In the end, two years later, Santer would win a MacArthur "genius" award for his work, and receive a hearty e-mail from one of the scientists who'd attacked him in 1996.

"Congratulations. You won the game," it would say.

"It wasn't a *game*," Santer would say, after his divorce had gone through, and as he waited to go into a hospital for one more round of stomach surgery, which had been going on for months.

"It wasn't a game."

———

One Saturday morning that same summer, another worried scientist, 30-year-old German Andreas Kääb, loaded up his four-wheel-drive Subaru and drove out of Zurich toward the village of Saas Balen high in the Swiss Alps. He was dressed for hiking, one of his favorite activities, but the closer he came to his destination the more anxious he became over a different sort of danger from global warming.

Kääb, a Zurich resident, studied mountain glaciers, usually from aerial photos and computer imagery, and he knew that a glacier above Saas Balen was rapidly melting, filling three drainage lakes below it with potentially life-threatening runoff.

"Ten years of warm weather was responsible."

The lakes were partially dammed by ice, but if the glacier melted too much the ice would rise, the dam would become a floating iceberg, and the waters from all three lakes would join into a torrent rushing toward the town.

To prevent a climate-caused tragedy Andreas's boss at the Technical University of Zurich, Dr. Wilfred Haeberli—also head of the World Glacier Monitoring Service—had received funding from a Swiss government program designed specifically to pinpoint potential mountain disasters from global warming. Kääb had studied the rising lakes from aerial shots and made computer projections of melting. Now it was time to see up close if the dams were due to break.

Summer was the time of greatest ice melt, greatest danger, so Kääb, whose mountain trips usually involved more pleasurable skiing or cycling, felt his anxiety quicken as he drove through Interlaken and loaded the car on a train to pass through a tunnel.

On less nervous days, he, Haeberli and the handful of researchers at Zurich's World Glacier Monitoring Service shared an easygoing attitude and passionate love of mountain glaciers that had started for Kääb and Haeberli in boyhood, separately, when they'd first walked in the Alps.

"It is a love story," Haeberli would tell reporters who frequently interviewed him, audiences at his lectures and government officials from all over the world who consulted him about disappearing glaciers. The lanky, mustached professor thought of glaciers as "individuals," each with a separate personality, each fascinatingly unique.

The Columbia Glacier of Alaska, for instance, the "Queen of Alaskan glaciers," he would say with reverence, showing slides, was "a beautiful, attractive queen. Everyone is attracted by her blue colors. By her crevasse patterns, which are wild. And yet she has a soft appearance. But she has been turned into a threatening queen because the front has receded and she has produced many icebergs which have floated into the Gulf of Valdez, and threaten the tanker lines moving oil from the Alaskan pipeline."

The Gruben Glacier, Kääb's destination today, was "smaller, on a very steep slope. More difficult to access, but very interesting and sometimes dangerous.

"When one stands on a glacier, you are on the back of an animal," Haeberli said.

And the animals were stirring. By 1996, after 100 years of monitoring glaciers, World Meteorological Organization researchers knew the majority of them were shrinking, and many were becoming dangerous. Glaciers on Mt. Kenya were losing mass and distance. So was Djankuat Glacier in Russia, Urumqihe Glacier in China, Peyto Glacier in the Canadian Rockies, Zongo Glacier in Bolivia, Vernagtferner Glacier in Austria. All were falling back.

In countries like China and Argentina, Haeberli would warn audiences, "the disappearance of glaciers will eliminate a source of water supply during summers. Glaciers provide drinking water and irrigation."

In parts of the Himalayas, South America and the Alps, shrinkage would pose more immediate danger from floods.

"Mountain glaciers are a clear indicator of global warming," Wilfred Haeberli added in speeches, because mountain glacier behavior is simple to understand.

"The formation of mountain ice is high school physics, understandable even to people who have never seen a glacier. Everyone has had their ice cream melt because it is too warm."

Also, glaciers kept records of their own behavior, Haeberli would inform fascinated audiences. "They *have a memory*." They leave striations in rocks marking their paths, and deposit debris when they stop. Tourists have taken photographs of glaciers, or painted them, for centuries, providing evidence of their shrinkage.

In fact, the only glaciers that *weren't* shrinking, Haeberli and Kääb knew, in North Atlantic areas like southwest Norway, were located in places where climate models predicted temperatures would *cool* as global warming altered ocean currents, forcing warm waters away from land.

But now, parking outside the only shop in Saas Balen, Andreas Kääb was concerned about dangerous melt flow. Much of what he saw around him would be submerged if the glacier melted too fast. The land was so steep that the town had no flat areas. Wildflowers abounded beside small neat homes of weather-darkened wood. Farmers cut hay by hand. A few cows grazed on sloping meadows.

The lanky, bearded German bought bread, cheese and sausage and filled a thermos with hot tea "because I will have to go up 3,000 meters and even in the summer it is cold." The shopkeeper, a heavyset man in an apron, recognized his visitor, and asked, "Are the lakes safe?"

"I'm going to find out."

The Subaru brought him up a sharply rising gravel road that terminated several hundred yards from the lowest lake, still out of view. Kääb started hiking. The glacier loomed ahead, hemmed in by mountains, its terminus covered with boulders from rock falls, its body of clear ice tinged gray-blue. Kääb passed the vegetative zone and entered a moonscape of glacial debris: rock, boulders and ice.

Then the first lake came into view, and he relaxed a little. The turquoise water, at least in this lowest lake, remained at a safely low level. Kääb sprawled on a rock in the sun and ate lunch. At least if melt danger existed here, it was not immediate, and would come from a higher lake.

Was there a connection between the global warming causing glaciers to melt and human-made greenhouse gasses? Sitting, chewing, the young German believed like Wilfred Haeberli that the glaciers themselves provided a link.

Haeberli put it this way. Although glaciers had *started* retreating 200 years ago, when human influence on the atmosphere was nonexistent, "the *rate* of retreat *can* be related to human impact . . . And the *rate* has accelerated to the point that . . . the amount of energy needed to shrink glaciers at the speed they have been shrinking, roughly three watts per square meter, is the *same* amount of energy allegedly put into the atmosphere by human-made emissions."

In other words, by the 1990s, if there were no human greenhouse emissions, *there would be no shrinkage.*

"For the first time in earth's history, glacier shrinkage may be predominantly influenced by human impact on the atmosphere," Haeberli said.

What frustrated him about his belief was the same thing that bothered Jim Hansen, Ben Santer and dozens of other scientists concerned that humans were altering climate. They could not *prove* it 100 percent and that was being used as a political reason to avoid the issue. Scientists had been given an impossible role. They were expected to know the future. But *no* scientist would say that anything was 100 percent sure without a test.

To put this in perspective, would NASA scientists have been able to say, positively, that the first moon rocket would work before they launched it? Would the men and women who worked on the Manhattan Project have been able to say, with absolute certainty, that the atom bomb would work before one went off?

Never. Yet policy had been based on their educated projections.

When it came to climate, though, natural scientific caution was instantly seized upon as an excuse to do nothing, and Haeberli hated it. "We should be more careful *because* we are aware of uncertainty. I feel as if we are in a car, and the speed is increasing. Ahead is a sharp turn, but we do not know how sharp. Some people say, let's brake a little. Others say, let's wait until we know the turn will be sharp. The car keeps accelerating. If we brake now we'll have the option of deciding at what speed we want to take the turn. Otherwise we will only be able to decide whether to keep our eyes open or closed during the accident."

Today Andreas Kääb's eyes were open as he tried to *prevent* an accident. He reached the second lake and, relieved, saw that the water level was fine.

But when he made it up to the third and biggest lake, he thought, *uh-oh.*

It was much higher than it should be. Chunks of loose ice floated in the water, and he spotted a large hole in the ice dam, 30 feet high, 90 feet wide.

He said, out loud, "They better drain the lake."

Back at the Subaru, he found the engineer who had been hired by the town to make sure the lakes did not overflow. The two men had arranged to meet here, and they spread maps on the car and considered diagrams of a drainage project the engineer had designed.

"You better start soon," Kääb advised.

The engineer did, and fortunately Saas Balen would finish the millennium without suffering a flood.

"But in general there are other high mountain lakes which could be dangerous in Switzerland," Kääb said, back in his Zurich office. "And globally, there are *huge* lakes in the Himalayas and South America, *much* more dangerous, in heavily populated areas. The likelihood of disasters will increase until glaciers reach some kind of new equilibrium with the climate. I wouldn't be surprised soon if I pick up the paper and read about whole communities being submerged in the Himalayas or South America. Thousands dead."

Abdullahi Majeed, diminutive and quietly outspoken chief meteorologist of the Maldives, deputy minister, representative to the World Meteorological Organization and brother of President Gayoom, listened to the American with growing disgust. Outside the cafeteria window of Geneva's Palais des Nations he could see gorgeous green meadows and trees. The 1996 climate summit was under way and both men, sipping tea, were part of the 1,500 delegates, journalists and NGO representatives crowded into the hall.

Majeed had been explaining how rainfall patterns had been altering in the Maldives over the last few years and how many islands that were only a few feet above sea level—populated ones—were losing land at an unprecedented rate, to ocean erosion.

That was why the Maldives was seeking $1.2 million in Geneva to fund a project to inventory and study ocean rise.

The American listened and tried a helpful suggestion.

"Why not join forces with other countries in the region and build sea walls?"

"There aren't any other countries in the region," Majeed replied.

"Oh," the American said, smiled after a pause, and resumed. "You know what I mean. *Other* island countries. Seychelles. Comoros."

"You're not living in the real world. They have their own budget problems."

Nearby, in the cafeteria, another forlorn presence, Dr. R. K. Pachuari, head of India's Tata energy institute, TERI, and a vice chairman of the IPCC, sat with friends, criticizing his own delegation. India's refusal to agree to mandatory energy cuts—and China's—was being used by treaty opponents as an argument to keep other countries from coming to an agreement. Pachuari felt India was handling the issue the wrong way.

It wasn't that he thought India should or *could* take on binding reductions, given its energy needs. But he also felt that "the impact of climate change will be overwhelmingly felt in the developing world . . . Ours is such a system of fragility," he would say, speaking for many poor countries, "that anything that upsets the balance will have enormous effects on human beings here."

That was one reason he felt India should be making it easier for other countries to sign a treaty. Instead of just saying "no" to cuts, India should be stressing its efforts to reduce air pollution back home, like the government's tax credit program for solar or wind power projects. That would show India was serious about curbing emissions. Hopefully some day, when India was more developed, it would join the treaty.

But at the moment India and China seemed only to favor technology transfer to poorer countries: aid, not targets.

Pachuari was tortured by a particularly frustrating view of these dynamics because he occupied a unique niche between the developed and developing worlds. He had lived in the U.S. and taught at U.S. universities. He enjoyed friendships with many Americans and knew their point of view. But as head of an institute dedicated to developing renewable and cleaner power for India, he also saw the gap between worlds.

The affable ex-academic was a slight man with longish gray hair and a quiet sense of humor, which might manifest itself by his wearing a respectable blue sports jacket with a Tweety cartoon bow tie. He had started his career as one of only eight students selected, from thousands tested, for India's most elite engineering school.

The love of nature came from boyhood, when he'd grown up in the mountains of northern India, son of an "educationalist" who had taken him on horseback trips to tour outlying schools. Wonderful journeys, where he'd catch glimpses of leopards and panthers. He'd never forget the night a tiger sat down among the porters as they cooked dinner on an open-air fire. The tiger had not attacked anyone. It had not eaten anything. It had simply sat while the porters realized it was there, and their singing, as they were too smart to flee, grew sad and afraid.

After a while, the tiger walked away.

Now those mountains, like much of India, were spoiled. The Himalaya glaciers were melting. The city air was foul. In New Delhi, Pachuari worked in a soup of exhaust-laden air, and he knew pollution had to be cut.

From TERI's beautiful red brick headquarters, a modern complex set beside a road clogged with three-wheel pedicabs, smoke-spewing taxis and even an occasional ridden horse or elephant going by, TERI promoted solar and wind power and tried to develop clean-emission technologies for India. It provided a meeting place for representatives from the Indian power industry, overseas corporations and foreign aid organizations. It sponsored seminars on energy for India's elite. Its staff—the cream of India's engineers—worked to design technologies tailored to a developing country—from wind turbines to brick kilns.

So now, in the Palais des Nations, Pachuari saw the stakes with particular clarity. But still he felt, "The U.S. shouldn't be pushing down our throats the need for reducing greenhouse gas emissions. The U.S. should be joining hands with us to improve the local environment. While you do that, you attack greenhouse emissions at the same time."

That's what he told Americans like the U.S. Senator who'd expressed astonishment when he learned how little energy an average Indian used compared to an average American.

"He asked, how do you manage? I said we have 750 million people who are not even part of the modern energy system," Pachuari said.

"We were in two different universes, I fear."

And as the Geneva conference began, it seemed to the depressed Dr. R. K. Pachuari that all parties in the "different universes" seemed worn down by six years of bickering. He feared the conference would collapse if the U.S. refused to budge on targets and timetables, as usual.

"The enthusiasm of Rio had withered away."

Outside the cafeteria, in the big meeting halls and small private conference rooms, hallways, hotels and restaurants, the same old arguments broke out as if there had been no hiatus. Fights erupted over rules of procedure, over who could attend meetings, over how to phrase differences of opinion, word by word.

Russia proposed that the full plenary session vote on ways the IPCC report should be used in policymaking, but Saudi Arabia blocked consensus. Representatives of oil-exporting countries stuck their heads in meeting rooms, during meetings, and called, "We object," even though they hadn't been there until then. The U.S. and Iran proposed an information clearinghouse for technology transfer. Syria argued that more research was needed.

For Pachuari, usually an upbeat man, "everyone was defensive. People didn't want to commit to anything."

It was like some gigantic, anarchistic, environmental tower of Babel. Even unhappy representatives of the insurance industry were present, having finally come to the conclusion that climate change was real and dangerous. A consortium of 54 international companies, including Swiss Re, AEGIS Insurance Company of South Africa, CIGNA of Japan and corporations from Germany, Russia, Argentina, Canada and Italy had released a statement urging "a substantial reduction in greenhouse gas emissions."

But if treaty proponents were feeling sour, so was their opposition. The campaign to discredit the IPCC had not worked, and with Bill Clinton up for re-election, energy companies feared he would pander to environmentalists and announce support for mandatory energy cuts. Plus there was always the maddening Vice President Gore to worry about.

"Gore was the most influential vice president ever in this country," said Bill O'Keefe, head of both the American Petroleum Institute and the GCC.

The companies had not been idle. On the day run-up meetings had begun for the conference, they'd sent Clinton a "Dear Mr. President" letter, signed by 119 CEOs and corporate chairmen, saying, "The U.S. should not agree to any of the three proposed protocols presently on the negotiating table. Your leadership on this issue is critical to a continuing strong U.S. economy."

Signatories included the heads of Amoco, Chrysler, Exxon, Mobil and Occidental, and the GCC's argument had changed now that efforts to discredit the science had failed. Treaty proposals on the table, the claim went, would hurt the U.S. by giving a leg up to competitors like India or China. With the U.S. hamstrung by new emission laws, India and China would steal away business.

Said R. K. Pachuari, "The whole process was grinding to a stop."

Round and round. For Pachuari, it was like listening to some torturous long-playing record, whose needle was wearing into the grooves.

But then something amazing happened near the end of the conference, when Tim Wirth got up to give the speech outlining the official U.S. position.

Pachuari listened with growing astonishment as Wirth announced that the U.S. had decided to *agree* to adopt legally binding targets. It was the hugest turnaround in six years. Pachuari grinned as boos broke out among U.S. conservatives.

Wirth was triumphant. It had taken eight years since Jim Hansen testified before his Senate committee to reach this point, and the speech represented the high point of how the U.S. was viewed internationally on the climate issue. The Clinton administration had actually agreed to take on legally binding measures to cut fossil fuel use in the most energy-hungry nation on earth, and it would do so next year, if things went as planned, at the 1997 Kyoto summit.

Pachuari returned to New Delhi, buoyed. "Without that speech, the whole thing would have ended with a whimper."

Geneva had been a giant turning point. But back in Washington, the two biggest problems for Wirth, Gore and Clinton were ones that Dr. Pachuari didn't know about.

The White House still had not determined how much energy could be cut from the economy without damaging it.

And even if the administration signed a treaty in 1997, the U.S. would never be bound by it unless the predominantly Republican Senate voted to go along.

Still, it was turning out to be a bad year for Bill O'Keefe's GCC, and now another enormous blow fell. In an extraordinary turnabout, the energy giant BP withdrew from the organization and announced that human-induced global warming might be real. The belief that the climate was deteriorating had finally reached a boardroom of big oil.

The astounding break in the formerly uniform position of fossil fuel companies had begun with a man named John Gore (no relation to Al) and a presentation he'd given to top BP officials earlier that year at global headquarters at One Finsbury Circus, in London's financial district.

Going into that meeting, in which he was to report on the status of the climate change issue, Gore, an American and former lobbyist for BP in Washington, had been fighting the climate treaty for years as company representative to the GCC and American Petroleum Institute. He was Group Vice President in charge of Government and Public Affairs, a kind of government affairs tactician.

Even after years of immersion in the issue, he had no opinion on the actual science of global warming.

"Facts? There are facts everywhere. I can give you any set of facts you want. You can find a scientist who says the end is near. You can find a scientist who says it's a load of crap. The facts had nothing to do with it."

Gore's area of expertise was strategy.

"The business I am in is the sale of ideas. You need to understand your customers. Governments. NGOs. Consumers. Shareholders. Politicians primarily. What will be acceptable to them? Believable? What will be something they want to hear and act on? And hopefully meet the larger needs of society? But my primary responsibility is to say what will sell. Tactically, it's what the world out there *thinks* and what they're going to *do* that I'm interested in."

Going into the meeting that day, Gore was enmeshed as usual in the strategies of both sides in the fight.

"From the green side, climate change was a wonderful thing because if you could get people convinced that the world was in danger because of climate, you could get an off-oil agenda you could otherwise never achieve. Climate change could be the lever to convince governments to do things, to get people to use less fossil fuels . . . Otherwise, you can't get people to stop driving cars. *Very* clever.

"And tactically," he would say, savoring the word, "*also* very clever, if you're going to fight climate change, you don't want to fight it as big oil. You want to have people who are friendly, popular, lovable, fighting it . . . Labor unions . . . Farmers . . . Auto companies . . . The GCC, as a front group, embraced all of these other lobbying forces."

And now, as the meeting began, the tall, fit-looking and well-dressed Gore, talking strategy, told CEO John Browne that, in his opinion, GCC tactics were failing.

"I rejected Bill O'Keefe's argument that we keep hammering the science as unproven . . . That you should fight tooth and nail in the trenches anything that concedes, even in the slightest degree, that this might be a problem.

"I thought a better approach would be to say the only way this thing will ever be solved is for a real dialogue to take place between developing and developed countries. Until the U.S., China and India agree on a way forward, all the stuff around the edges is meaningless."

Of course, Gore hadn't the slightest idea whether there even *was* a climate problem. "I knew there was a *perceived* problem. The sale of that idea was progressing faster than the sale of the idea that there was no problem."

But when Gore finished his presentation, John Browne asked him a question that, in all his years in climate politics, he had never heard from a boss.

"*Is* there a climate problem?"

Gore stood there, astounded. Then he said, "You want to go into the *science?*"

Yes, Browne said. He wanted to understand the actual science. So Gore put together a "top-notch panel of scientists" and collected scientific papers, which people like Browne and Rodney Chase, the number two power in the company, reviewed. The process took "a couple of months." Browne met the scientists, talked to them.

Browne and Chase came to the same conclusion, that "there was sufficient case being made for us to have concerns that we may be part of a phenomenon, of changes on earth," Chase said later, in a more aristocratic British style contrasting with Gore's rapid American PR-man delivery.

"It was not possible to prove it. It was certainly possible for some to argue vehemently against it," added Chase, a well-dressed, conservative, blue-eyed power in the industry. "We felt that the science we had investigated was of sufficient concern for us to believe we needed to be part of an active response."

So, behind the scenes, BP executives like Chase began having "active disagreements" with counterparts at other oil companies. He was a member of a culture valuing the corporate good over that of the individual—even himself—and Chase tended to be more comfortable explaining the new BP view in terms of tactics rather than beliefs about climate . . . at first.

"Conversations about concerns on global warming within the professional oil industry were always full of reluctance," he said.

"BP was always alone in these discussions. We talked for example about conservation. Should we be activists as energy leaders in proposing solutions for ameliorating climatic impacts? The total level of energy consumption? The nature of greenhouse emissions? The pace at which technologies on which corrective actions could be based? *These* are the issues we talked about in industry forums, and our position was unwelcome."

Chase himself spent time in the U.S. and Europe arguing with "members of the automobile industry on strategic and research-based programs to set up improvements in the way energy is used in transport. We [BP] said we wanted the joint research programs to begin to focus on the speed of response to emerging climate concerns, on improving emissions consequences from motor vehicles . . . Their arguments were always couched in the same way. That the science of climate change is unproven. That it is a political issue. That it was too soon to be taking precautions. That environmental advantages don't just exist in a vacuum. They have to exist beside economic activity and well-being. These were not foolish arguments but we had come to a different position on priorities."

But the "anxiety" BP executives felt over the climate issue, Chase added cautiously, only after being drawn out, also reflected personal as well as tactical considerations.

"Actions are taken by individuals, not organizations. And I think individuals always need a trigger of some sort. For some it's intellectual. For some, emotional."

For Chase himself, it had been a combination of the two, not just scientific reports but real weather events that affected his view.

"I don't want this to sound artificial. The sort of events included incidents of weather change that were being reported in the newspapers. Hurricanes in some parts of the world. Floods in other places. In our country we had hurricanes, which had never hit our nation before, hitting us two or three years in a row. Storms of such severity that large areas of forest were damaged. People's homes were damaged," he said, referring among other storms to the one that had killed Emily McDonald in 1990.

Rodney Chase was in a different profession and occupied a different social stratum than the people in Grange Junior School who had watched the roof fall in that day, but he belonged to the same society, shared sensitivities, and used the same words to describe the storm that Bill Collins the fireman and Malcolm Emery the headmaster had used.

"These are things we read about in other parts of the world but they *never happen here*."

Chase added that his anxiety had also been partly a product of "personal events. Very personal. I suspect they go on in everybody's life."

The dinner story, for instance, had happened in 1995, at Chase's home in central London, where he'd been dining with his wife, son and daughter, a recent graduate of Bristol University.

Talk at the table had turned to the greenhouse effect.

"My daughter was discussing the role of energy companies impacting upon the environment. I don't think she realized, and I want to express this in a sufficiently humble way, what seniority I had achieved in one of the world's largest energy companies. At one point she said to me, 'Dad, I'm ashamed of you for not using the clear influence that you have to move your company in a different direction . . . Why would you not *do* that? Why would you not *use* the influence you have to make things better?' "

Chase felt rotten about it.

"It was a jolt."

He began telling the story around BP, in his understated upper-class way.

"I simply wanted to point out to my staff, look, we are all human beings, influenced by our surroundings, and if they wanted to understand where I was coming from, I not only needed to explain to them the intellectual basis of my analysis, but also the more visceral elements.

I remember thinking about that dinner many times, thinking, if I *do* believe that the environment is in need of more care, if I *do* think it is possible that our industry may be contributing to climatic change . . . then why wouldn't I use my personal position more forcefully?"

And so, by the fall of 1996, for a mix of scientific, strategic and personal reasons, BP executives had decided that their climate concerns and tactical considerations dovetailed, and after leaving the GCC they set about changing policy somewhat to reflect the new position. They supplemented oil and gas investment with money put into solar energy projects. They instituted an energy-cutting plan within the company, setting targets and timetables for different divisions to reduce consumption.

"Meeting these targets could make or break careers."

To his vast surprise, Chase found that the policy struck a huge chord inside the company. "It unlocked tens of thousands of emotional reactions in our workforce. We didn't expect it to. But it became obvious that this is what people in the firm wanted us to do."

And later, when BP took over the U.S. oil company Amoco, Chase, in charge of all U.S. operations, toured BP facilities there to learn employee concerns over the merger.

"I thought they would be worried about jobs, but environment was the only issue. I remember going out to a BP production platform off the coast of Mississippi. The workers told me, 'we like the acquisition, but we want to be sure it won't change BP's environmental posture.' Hard-nosed roughnecks, on platforms out on the deep waters of the Gulf of Mexico, and they said, *don't you dare backpedal on this policy*. Because Amoco was not known to us for having a similar position.

"It was extraordinary to me."

He said also that when it came to the Amoco executives, after the acquisition, he found himself in frequent "angry exchanges" with them about climate change policy.

Some American executives were furious over BP's stance, an understandable reaction, Chase said, as they had worked in a different environment. But Chase told the Americans they were working for a company with a different philosophy now.

"We parted company with people who preferred to work in a different way."

In a ratcheting up of Tim Wirth's efforts to change U.S. policy, around the time that BP pulled out of the Global Climate Coalition, early one weekday morning in the fall of 1996, Ev Ehrlich, U.S. Under Secretary of Commerce for Economic Affairs, walked into the Old Ebbett Grill, a restaurant near the White House, where Rafe Pomerance, Deputy Assistant Secretary of State for Environment and Development, waited at a table.

"We have a problem," Rafe told Ev.

Ehrlich was a glib, liberally minded economist with great affection for Tim Wirth, who had been somewhat his patron when he first came to Washington. A likable mix of academic and street guy, he could sprinkle technical economics with profanity, literary metaphor and wide knowledge of how practical government works.

Ev ordered oatmeal with no milk and a sliced banana. Rafe had eggs. They got down to business.

Tim needs help, Rafe said.

Now that the U.S. had committed itself to embracing mandatory energy cuts without having any idea what the cuts would be, several competing government studies were floating around Washington, predicting how easy or costly a transition might be. As usual, the fight came down to environment versus economy.

Rafe said, "The EPA and DOE guys are working at cross-purposes. Treasury doesn't like this, and we aren't sure about the Council of Economic Advisers."

Someone was going to have to force all these competing agencies to come up with one agreed-upon report, to tinker with the different models, define an analytical process and produce a study the opposing groups could endorse.

Otherwise, "The White House wasn't going to bother with a bunch of pissants saying, we disagree," Ehrlich said.

Pomerance asked Ehrlich if he would take the job. Other senior economic officials were too busy, or unqualified, Rafe said, and Ehrlich took the word "qualified" to mean that the other candidates "were economists, not ass kickers. I was an under secretary. I understood there's both a carrot and a stick. I understood how to impose discipline on a bureaucratic process and not be coy about it."

o It was a job he loved. Ehrlich was a chief statistician for the U.S.. "Ten thousand people worked for me." His office was where the numbers started that eventually led to billions of dollars allocated or withdrawn from states, cities, federal programs.

"Being an under secretary is great. You get to learn and do, the two best things in the world."

What he was learning and doing at the moment was the U.S. census, a task that left little time for worrying about climate.

"I looked at Rafe and said I'm too busy. The census is a political lightning rod. Racists want to preserve the undercount in my most important program. A constitutional ritual! It's the only way people of color get political power and federal funding allocated to them. Just by turning up the numbers! And that's my *day job*! . . . And I was also in a long-term program to improve the quality of GDP measures. The economic statistics. My organization made 'em! We had the first comprehensive review of those statistics going on in forty years, creating a long-term plan to improve them!"

But Pomerance kept pushing. Wirth needed Ehrlich.

Ehrlich began to weaken. "Tim deserved a break."

Pomerance pushed more.

Ehrlich thought, it's the right thing to do. Rafe is right. No one else will do it if I won't.

Ehrlich felt a stirring of excitement.

This is a Nathan Hale moment, he thought.

It was one of those pivotal crossroads to Ehrlich, when fate offers a chance to accomplish something meaningful—in Nathan Hale's case, spying against the British army during the American Revolution. In Ev Ehrlich's case, affecting the outcome of a global-climate-change debate.

Of course, Nathan Hale, failing in his particular mission, was hung by the British. Ev faced less serious consequences if things didn't work out.

Nevertheless, the principle seemed similar, not to mention, if Ev worked on climate, he'd get to spend time with the environmental engineering crowd, people he liked. He'd probably have lots of interesting conversations with his wife, who had a PhD in environmental engineering.

Ehrlich decided, "This would be cool."

So on the conditions that the Council of Economic Advisers would

232 · BOB REISS

support his study (which it did) and that "this wouldn't be a put-up job.
Wherever the numbers fall is where it's going to go," he said yes, went
to his office and told his political and press advisers ("Young guys, very
sharp. If I worried about political pitfalls, I'd never get anything done")
what he had agreed to do.

"*You did what?*" they said, aghast, knowing that climate change politics brought nothing but problems.

But Ev began his examination of other studies, which would give a
pretty good idea of the limitations of using statistics to predict the
future.

First he held a series of interagency task force meetings with the
"boys and girls from all the agencies" to learn what they were doing.

"The meetings draw these guys like shit draws flies. EPA and DOE
were there. State. Labor. Justice. If one agency is going to be there, they
all have to be there."

Ehrlich listened to the arguing economic modelers with various
degrees of skepticism.

"The best way to talk about economists and their models is to think
of Bill Sykes in *Oliver Twist*. He kills his girlfriend, stabs her, goes to
court and *blames the knife*. Economists get an answer and they say, the
model gave it to me. A Pilate couldn't wash his hands more quickly."

To Ehrlich, models were only as good as the suppositions going into
them, basic human beliefs upon which starting numbers are based, and
sitting in meetings with the modelers, Ehrlich decided that some
awfully unrealistic assumptions were behind the very different results.

For instance, a Department of Energy model predicted immense
shocks to the U.S. economy if energy taxes were instituted. "But it
assumed that if you announce that ten years from now, you'll need com-
pliance, and you're going to phase it in, people won't do anything about
it until the last minute. They assumed everyone would just wake up and
get hit on the head at one A.M. No one would prepare."

Stupid assumption, Ev thought.

The Environmental Protection Agency model, on the other hand,
predicted a very smooth transition, except "they were taking a picture
of the economy every five years instead of every quarter. That model
could glide past short-term disruptions. And they assumed things like
because it's possible to construct a hybrid car that gets eighty miles a

gallon, it will permeate the market in five years . . . They assumed there would be huge proceeds from the sale of energy permits, and they would be pumped back into the economy like a dose of methadone, getting the economy to look good."

Bad assumptions, Ehrlich thought.

"I became kind of an intellectual and analytical traffic cop . . . I'd tell 'em, cut the bullshit! Let's make the assumptions transparent, decide which ones are right. Which models complement each other . . . I'd put the guys in front of me and grill 'em. What are your energy efficiency improvement parameters? How did you estimate them? Over what period? How does your model deal with utility boiler choice? What technological options does your model recognize for space heating and cooling, and over what increments?"

Ev relished the process.

"I'd been around energy models. I'd built energy models. I knew shit from shinola here."

As Ehrlich worked, Tim Wirth and Eileen Claussen began spreading the news among senators—potential friends or foes if the Senate was asked to ratify a treaty in 1997—that a special study was under way that would pinpoint, finally, the economic impact of transition.

Analytic squabbling would no longer hold up the process, Ehrlich felt, but he would turn out to be wrong. His study would only complicate things for Tim Wirth.

Meanwhile, 1996 would turn out to be the fifth hottest year on record. There was no doubt about the *heat* numbers. The 1990s were becoming the hottest *decade* on record, with the 1980s number two.

Washington ... Kyoto

WAITING FOR THE MAN from the CIA to show up—the caller who had told Richard Matthew he wanted to meet "as soon as possible"—the Georgetown University professor felt intrigued, curious and a bit nervous.

"I wondered what a CIA man would look like. Also, there's a myth in my field that if you work for the CIA, it's harder to get into other countries. What if I worked for the CIA and the next time I wanted to go to Pakistan, they wouldn't let me in?"

It was July 1997, another unusually hot summer, and the temperature—over 100 degrees at noon—made Matthew think about climate changing. After all, climate was part of his odd field of expertise. His bookshelves were filled with studies of environmental stress and its connections to war, revolution and state failure.

In his seventh-floor office in the University's Intercultural Center, a room devoid of posters or music ("I never like knickknacks in my office"), the mild-mannered, self-deprecating academic found himself thinking back to Princeton grad school: '86 to '90, when CIA recruiters came to campus and, "No one showed up for an interview. At least nobody said they did."

He also remembered how his adviser back then had said Matthew was making a big career mistake by focusing on links between environment and destabilization. No one would be interested in such a topic, the adviser was sure.

After the call from the CIA, "I was feeling a bit vindicated," Matthew said.

He wasn't exactly sure how Norm Kahn, the name the caller had given, had gotten his number. Maybe someone in Al Gore's office had passed it along, since recently Gore's staff had phoned to request copies of Matthew's work.

Or perhaps someone from the CIA had heard him speak at the Woodrow Wilson Center in Washington, on global environmental security, or heard his talk to the Defense Intelligence Agency in 1996, at a meeting in the Sheraton Hotel in Crystal City. *That* day had been bizarre as far as Matthew was concerned. First he'd had to pass through a security checkpoint just to get into a hotel room. Then lots of serious-looking people had been "walking around and asking questions, not explaining who they were. You have to guess who they are," he laughed. "People not giving their cards to you, which meant they weren't academics. Academics *always* hand everyone our cards. People asking you to stay in touch, but they wouldn't tell you how to do it. The attitude was, we'll call you; never mind you call us."

Now, at 1:30, came a knock, and Kahn, entering, "seemed like an average-looking, middle-aged white guy in a summer suit; my preconceived notion of a government worker. Short gray hair. Friendly. Not memorable in appearance, but he had a very quiet voice."

Matthew was a bespectacled, brown-haired "tie-and-jacket kind of guy, who loses the tie in the summer." His habit of dressing modestly, to blend in, came from regularly traveling in the third world, to get a first-hand look at countries he studied, and where he feared that standing out would give him a better chance of getting robbed.

He dated his fascination with climate to his boyhood in rural Quebec, where frigid winters had controlled every aspect of life, from starting cars in the morning to what sports he played.

Kahn got down to business.

He was, he said, a senior analyst with the CIA's Environmental Center, which drew on experts outside the agency and worked with other government departments—including the State Department, Department of Defense and National Security Council—to examine security

aspects of global environmental issues, like prospects for treaty negotiations and environmental crime. The center also analyzed potential security problems linked to environment. The CIA expected to be called
upon more in the future to advise the government on how these issues
affected U.S. security, economy and defense.

That made sense to Matthew. "A growing number of experts
thought we are at the beginnng of a cycle where fairly serious changes
in the environment are going to start driving a lot of conflict.

"With the cold war over, there was a debate going on how to
restructure U.S. security policy. Some people thought terrorists would
be the big threat now. Some people thought trans-national criminal
organizations. Some people thought security and environment. Al Gore
had put some pressure on the Defense Department to look at it aggressively, and there was receptivity there. Over the last couple of years there
had been a range of studies . . ."

Over coffee, the CIA man asked if Matthew would identify specific
regions where problems might arise, and Matthew considered various
environmental stresses that could push a potential trouble spot toward
eruption. Overpopulation. Toxic waste. Deteriorating nuclear reactors.

"Climate change was definitely one element. For some, *the* most
important element. For me, if worst-case climate scenarios turn out to
be true, an important threat."

He told his note-taking guest, "When people talk about climate
change as a security threat they're often thinking about water shortages.
Places where there are dams. Turkey. Iraq. Syria . . . Or about low-lying
areas, like Bangladesh, or small island states vulnerable to ocean flooding. The situation could overwhelm the national infrastructure. You'd
have unemployment. Disease. Intensifying ethnic clashes. Rivalry over
the little remaining scraps of arable land and fresh water. A spiraling
effect . . ."

Too much rain, Matthew said, would cause floods to ravage already
shaky third world areas. He'd seen it, up close, in Pakistan and
Bangladesh.

Too *little* rain and drought could push competing countries or ethnic
groups toward war.

One Canadian Defense Department study Matthew had read, "Climate Warming, Water Resources and Geopolitical Conflict," had

already predicted that evaporation caused by climate change could exacerbate tensions in the Mideast; on the Jordan River, separating Israel and Jordan; the Nile, whose waters fed Egypt, Ethiopia and Sudan; and the Litani River in Lebanon, coveted by Lebanon, Syria and Israel.

"The continued decline in water levels in the Nile could threaten political stability . . . The next war in the Middle East, according to Egyptian officials, may very well be over water," the report said.

In short, climate change worsened an already bad mix of elements in places where the U.S. might one day bring aid, keep peace or fight.

"What you have in the developing world is millions of people living on the threshold of poverty, with very little to fall back on," Matthew said. "With globalization already upsetting traditional borders—already causing breakdowns of traditional societies, if the environment becomes one more shock, what are their choices? To leave? To sit in a pool of unemployment and disease and eke out an existence? Or to try to reach developed countries?"

These things had already happened in some places.

"The military is being asked to participate more and more in operations other than war . . . stopping boatloads of people coming in . . . going into humanitarian situations where it's very hard to identify the enemy because it's chaos. It's hard to know who the good guys are and who the bad guys are because what you really have is a lot of loosely organized gangs competing for resources."

Matthew went on to tell Kahn that in contrast to, say, California, "where . . . there are fairly good housing codes, regulated for earthquakes and flooding, elsewhere in the world—Bangladesh, Somalia, Egypt, Central America—people are taking their chances. They're not planning for big events. The hope is that the probability of experiencing big events is small, but climate change suggests that the probability will increase dramatically. Securitywise, people will be devastated. Look at Egypt in 1992. An earthquake hit and the government couldn't respond. Look what happened. Another force moved in. In this case, more extreme Islamic elements. A breakdown in central authority . . . Look at Central Africa. If the government can't do anything, up pop the independent militaries. Extremists. It makes for highly volatile situations."

Kahn asked Matthew if he would be interested in attending a "flash points workshop," which the CIA would sponsor that November, in the

Sheraton Reston Hotel in Virginia. A "flash point" was a place where environmental stresses might trigger conflict. Experts at the workshop would consider specific areas where that might occur.

"Climate change was considered a big driver of flash points," Matthew said.

Matthew told Kahn he would attend. Then he went home to his campus apartment and told his wife, with some pride, that the CIA had consulted him.

"Don't let it go to your head. You're not that important," she replied.

"We are dealing with an issue that has potentially drastic consequences for American foreign, domestic and economic policy," warned Senator Chuck Hagel, today's antagonist for Tim Wirth. "The course of action we take will affect for future generations our economy, environment, energy costs, economic growth, trade, jobs, global competitiveness, national defense and perhaps, most importantly, our national sovereignty."

The "issue" was the treaty taking shape for the Kyoto summit, five months away. It was 9:30 A.M., July 19, in the Dirksen Senate Office Building, and Hagel, a Nebraska Republican, chaired the Subcommittee on International Economic Policy, Export and Trade Promotion, of the powerful Senate Foreign Relations Committee.

Senate opposition was growing to the kind of treaty that last year's summit had set the stage for—one including binding emissions reductions for the U.S. while excluding developing countries from the same regulations.

Hagel and Senator Robert Byrd of West Virginia were so incensed that they were sponsoring a nonbinding resolution, to be voted on soon by the Senate, detailing conditions that would prevent ratification of a climate treaty.

There would be no ratification unless developing countries were included. No ratification if a treaty might cause serious economic harm to the U.S. No approval unless the administration presented the Senate with a list of costs, laws, legislation and regulatory action that would be required under a treaty.

The resolution had already received the support of 61 senators, enough to kill any treaty, which meant that Tim Wirth and Assistant

Secretary of State Eileen Claussen, chief negotiators, were locked in a three-front war as Kyoto approached. In the Senate they defended the White House position. At the White House they battled to *figure out* the position. In runup negotiations to craft a treaty, they fought other diplomats over what provisions it would include, but their efforts were hampered by their inability to define U.S. goals.

"If you're going to conduct a successful negotiation you have to be clear about what you want in the end," said Claussen. "It was very difficult through the first six months of 1997 to push the White House to focus on it. There were other issues on the agenda . . . Virtually every negotiation I'd go to, people would ask me, what kind of targets is the U.S. willing to accept? Well, it's hard to make a case for all the other things you want if you can't talk about this, because in the end it's all a trade-off."

So Claussen would "either make something up, or say we don't have a number yet."

Meanwhile, today's Senate grilling would be representative of the frustration Wirth would endure for months.

The tall, lanky Coloradan had the flu, but still opened his statement with soothing praise for the senators. He was pleased to join them and "especially pleased to follow this morning two gentlemen I consider important mentors to me," Senator Robert Byrd and Congressman John Dingell of Michigan, but both men were here on the attack.

Wirth smoothly went over the IPCC report and its predictions of disruption from global warming. He produced a statement signed by 2,409 scientists urging a clear national U.S. plan to limit emissions. He said U.S. negotiators were trying to establish a three-stepped binding emissions policy for developing countries; but he stressed that without mandatory targets, *developed* countries would merely "pay lip service to their efforts to solve this problem."

Wirth spoke passionately but inside he was growing resentful toward the White House. Bill Cinton and Al Gore gave good speeches about global warming but from Wirth's perspective they were not following through with support.

"What did they ever do to push this policy? Not much."

The White House staff "wouldn't even show up for briefing meetings," Wirth would recall.

⟂ Not to mention, Ev Ehrlich had finished his study of the economic consequences of energy reductions, which Wirth and Claussen had promised to show the Senate, and the White House was sitting on it. The study showed that yes, there would be some costs in cutting fossil fuels and no, the costs would not be catastrophic.

"The White House didn't like it," Wirth said.

"It was absurd," Claussen said. "How do you build a consensus if you're not willing to show these numbers?"

But Katie McGinty, coordinator of White House environmental policy, thought Wirth's position unrealistic.

"There was no sense in being self-righteous and talking about things that you know are not reachable," she said. "That is the height of disingenuousness, the kind of posture that does not lead to actually addressing an issue . . . We had to be steering a course that we felt, one, could attract the support of other countries to help congeal an international agreement and, two, actually lead to emissions reductions. It had to be a policy that was economically and technologically feasible."

To McGinty, even the Byrd–Hagel resolution, which certainly *looked* like an enormous repudiation of the administration, contained loophole language that could enable the Senate to ratify a treaty. The resolution did not demand that developing countries take on the *same* commitments as developed ones, so perhaps if U.S. negotiators could convince countries like India and China to accept *any* commitments, the Senate would eventually go along.

Meanwhile, in Dirksen room 419, Tim Wirth was in the hot seat in the building where he used to chair hearings.

Senator Byrd produced a chart showing projected carbon dioxide emissions in the year 2015, if the proposed treaty went through. In the chart, a whopping 50 percent of emissions came from the *developing* world, 34 percent from the U.S. and western Europe, and 14.5 percent from the former Soviet bloc. Byrd said that if the proposed treaty did not equally commit developing nations to binding commitments, there would be no incentive for them to make environmentally sound choices, and "no incentive for the Senate to approve such a treaty. I can guarantee you that there will be a mountain in the way which a mustard seed of faith will not . . . remove."

Congressman John Dingell of Michigan warned, "Are we setting the U.S. up for economic fiasco? . . . It is very clear that the State Department has either been rushing forward without information, has been suppressing (Ev Erhlich's study), has not been making it available to Congress . . ."

It was brutal for Wirth. Pat Michaels debunked the IPCC science. A parade of business and union officials from the AFL-CIO, National Association of Manufacturers and Nebraska Farm Bureau Federation predicted massive job losses and farm closings if the U.S. took on *any* binding emission targets.

Wirth stressed costs if climate change was *not* addressed. It wasn't as if doing nothing would be painless. A 20-inch rise in sea level would inundate 9,000 square miles of U.S. coastal land, he said. And heat waves "like the one that killed 500 people in Chicago two summers ago would be four to six times more likely to occur."

Hagel pressed Wirth, asking if his ultimate goal was a whopping 70 percent emissions reduction of greenhouse gasses.

The tough bottom-line question addressed the dirty little secret of fuel-cut advocates, since climate researchers agreed that cuts being proposed for Kyoto—5–20 percent—*would not do the job in the end*. Wirth had to admit that if "you were to double the pre-industrial age amount of CO_2 in the atmosphere," and if that "is the ultimate concentration we believe we can tolerate," then achieving safe levels of CO_2 "will require, ultimately, a seventy percent reduction."

Hagel let that daunting number sit there. Seventy percent! Then he reminded Wirth that Chinese officials had consistently said they would never take on reductions, and Wirth countered that, sure, they said it, but in negotiations, lots of times, people say one thing in the beginning, as a bargaining position, and another at the end.

It was a bad morning for Wirth, yet there were glimmers of support. Maryland congressman Wayne T. Gilchrest, whose district lay on the Atlantic Coast, submitted a statement saying that to the people of his district, "the cost of doing nothing is too much to bear."

The Senate passed the Byrd–Hagel resolution 95–0.

"I think my frustration was equal to Tim Wirth's," Eileen Claussen said.

Her biggest problem with the resolution was that it required developing countries to take on targets *in the same time frame* as the U.S.

"As a practical matter, you're not going to get China or India to agree . . . When Senator Byrd was walking around the Senate, door to door . . . asking people to sign the resolution, it should have been possible for the administration to say, you can't solve this problem without dealing with the developing world, but let's work on language that recognizes the principles we agreed to in Rio, something that says they have to be involved but doesn't function *in the same time frame*. Quite honestly, the administration refused to take on Senator Byrd . . . Part of that, I thought, was an unwillingness to put a huge effort into something that was going to pass anyway . . . But is it fair to expect a country whose per capita emissions are one tenth of yours, whose GDP per capita is almost nothing next to yours, to do the same thing as you?

"In the end I thought that was so terrible. That was it for me."

Claussen quit.

Wirth stayed, embattled. He would always remember one stinging comment from one top Treasury official after a briefing on Capitol Hill.

"He said, 'You've gone native, haven't you?'" meaning "that I'd gone over to the climate side of things and wasn't representing the administration's interests."

The breaking point came after Wirth went to dinner one night with Atlanta billionaire Ted Turner. Turner offered Wirth a job running his new "United Nations Foundation," an organization being established to execute Turner's $1 billion pledge to support UN economic, environmental, social and humanitarian causes. Wirth said he'd think about it, and reported the offer to the White House Ethics Office.

"The White House immediately decided to take me off the case. They said there was a conflict of interest, but there wasn't . . . They were using it as an excuse to get the guy who had gone native out of the deal.

"I wasn't going to make a big stink about it. Hell, it was pretty distasteful working with these people anyway."

Fired, Wirth left with two weeks to go before Kyoto. He would take the job with Ted Turner and the U.S. would field a new negotiating team in Japan.

As the announcement went out that 1997 was one of the three

hottest years on record, the full-scale lobbying campaign against the treaty continued in the U.S.

Would the climate treaty become an environmental League of Nations, a lost opportunity and public regret?

Or would it turn into one of those historic misjudgments, when a powerful nation hands its competitors power, setting the stage for its own later problems?

As December began, thousands of delegates, NGOs and journalists prepared to descend on Kyoto.

Chuck Hagel would go, determined to use every opportunity to stress that the "administration is not the solo voice of the United States," and that the U.S. "will not move forward unless it had the Senate, and it clearly did not have the Senate," a Hagel aide said.

Al Gore wasn't even sure he would attend. Like George Bush before Rio, he feared embarrassment if the process fell apart.

Tim Wirth stayed home.

———

The riots started in the northern cities. Then the demonstrations spread and the dying began. Troops manned intersections. Refugees clogged border checkpoints. Armed insurrectionists fought police.

Richard Matthew watched the broadcast, fascinated, sipping hot coffee and chewing a Dunkin' Donut. Around him sat dozens of military officials from NATO countries, and a U.S. Principal Assistant Deputy Under Secretary for Defense named Gary Vest. In autumn 1997, Vest had ordered the reality-based war game held. His specialty was environmental security.

They were in Carlisle, Pennsylvania, at the Army War College—a converted Indian School which included a military history institute and a "Center for Strategic Leadership," a three-story brick building crammed with the latest in information learning technology beneath its green tin roof.

War game participants sat in a third-floor conference room resembling a small version of the UN General Assembly, with booths for translators, a podium, stage and media screens for teleconferencing. At a horseshoe-shaped table in the front sat Gary Vest and co-chair Kurt Lietzmann, of the German Federal Ministry of Environment, Nature Conservation and Nuclear Safety.

The 60-odd participants watched a crisp CNN-style broadcaster give "news" from Eastern Europe. In the country of "Saphire" (based on Romania), the environment had deteriorated steadily since the fall of communism, crippling the economy to the point that mass violence was breaking out. Polluted rivers were devoid of fish. Heavy metals had contaminated crops and livestock and caused infant mortality to sky-rocket. Acid rain had destroyed Saphire's once rich forests. Saphire's food exports were banned across Europe as unsafe.

Now a leak had occurred at an aging nuclear plant.

Around Matthew, taking notes, were people who would have to deal with this kind of situation if it ever really erupted: Petr Kozel from the Ministry of Defense, Czech Republic; Risto Rautiainen from Finland's Ministry of Foreign Affairs; Gunnar Arbman from the Swedish National Defense Research Establishment; Major Volker Quante of the German armed forces . . . as well as grim-faced repre-sentatives from France, Poland, Hungary, Turkey, Lithuania, Switzer-land and the U.S.

As riots spread inside Saphire, the neighboring countries of "Glen-tana" and "Ronan" began military maneuvers near the border.

The biggest concern was "Rudyard" (Russia), to the north. Rud-yard authorities worried that their own nationals, fleeing home from Saphire, could destabilize Rudyard.

Would war break out? Should NATO send troops? How would powerful Rudyard react if *that* happened?

When the broadcast ended, Matthew and other role players were divided into five teams, representing NATO and the involved countries. They spent the day negotiating, guessing what the other teams would do, trying to defuse a kind of conflict they all knew could arise in Europe soon.

"Things got tense," said Kent Butts, the game's designer, a professor of political-military gaming at the U.S. Army War College. "People got into it."

Climate change was not yet part of the exercise, but when negotia-tions were through, Butts asked everyone to stop role-playing and to name and rank *real* environmental threats they expected NATO to encounter in the near future.

He was surprised when climate change came up.

He was *very* surprised when it ranked third of 22 problems.

"A substantial number of the military people put climate change at the very top," Butts said.

They apparently considered climate change a more immediate global threat than mass human migration, nuclear waste disposal, energy scarcity, natural disasters, land mines, AIDS, loss of biodiversity and air pollution.

Only nuclear accidents and water scarcity were voted as more pressing problems, and since water scarcity related to evaporation, climate change was involved in that too.

Certainly Richard Matthew and Gary Vest considered it a serious issue. For them, diseases like malaria, yellow fever and plague came to mind when the topic arose. "There's a good amount of evidence that one result of continued global warming will be a change in the location and incidence of vector-borne diseases," said Vest, a compact, carefully spoken man bearing a resemblance to former NFL coach Jimmy Johnson.

"Vector-borne diseases" were a danger that, to Vest, could have U.S. soldiers running high fevers, packing field hospitals and dying in foreign lands.

"Some authorities say, in the history of warfare, certainly modern warfare, there have been four times as many combatants rendered non-combatant status from vector-borne diseases than from the enemy. So if you're thinking in terms of security you pay a lot of attention to vector-borne disease. But you also understand that an outbreak can be destabilizing . . . At what point does the death rate become *so* destabilizing in a country that it produces violence in people, anti-government behavior?"

Environmental concerns were growing so much in the military that, by 2000, a country-by-country environmental assessment would be proceeding in the Pentagon, to identify potential trouble spots.

But even in 1997 climate change presented another, more immediate threat to the armed forces, and it came from the Kyoto protocol. Any attempt to control energy use in the U.S. would affect the Defense Department, a huge user of fuels. One of Vest's jobs was to help make sure military needs would be met if a treaty were signed.

"If and when a treaty requires implementation," he said, "what does that mean to us, as the organization that operates more cities (bases) than anyone else. Has more airports than anyone else in the world. Has more airplanes, more ships, more tanks. And is one of the largest consumers of fuel."

Meaning, if we have to cut energy use, can we still defend the country?

Chuck Hagel's office was already citing defense interests as a reason to oppose the treaty. Did American citizens really want UN clerks traveling with the U.S. Marines, toting up gasoline usage, gazing down at fuel gauges on airplanes, scrutinizing oil levels on battleships, applying every damn gallon toward some mandated national fuel cap?

"Next time NATO calls for a military activity," Hagel aides asked reporters, "will global carbon counters come along? Living under the Kyoto protocol, when we'd have to divvy up all our emissions, and count them against our domestic cap, those are the kinds of things we'd have to deal with."

But Gary Vest countered that the Defense Department, consulted all along the line, was satisfied with the U.S. treaty position. The military would send two representatives to Kyoto with the U.S. delegation—an admiral selected by the joint chiefs of staff, and an air force lieutenant colonel, to make sure defense needs were taken into account.

When it came to the military and climate-related problems, Vest worried more about vector-borne diseases.

"Or anything that produces mass migration," he said.

It was warm in Kyoto even in December. Autumn had not given way to winter, and trees in the ancient city dazzled summit delegates with their crisp yellow and orange leaves.

Inside the international conference center, on the last night, no agreement had yet been reached. The place was packed with over 1,500 government delegates, 3,500 journalists and 3,600 observers from environmental or business NGOs and government agencies around the world.

Katie McGinty was exhausted. Unless a breakthrough occurred, at sunup the Japanese translators would go home, workers would begin dismantling facilities and diplomats would stream home to report, *"THE CONFERENCE FAILED."*

She'd flown to Japan with Al Gore three days ago, part of his attempt to unblock negotiations. She'd stayed to join the negotiating team after Gore, instructing U.S. staff to "show flexibility," had returned home.

McGinty had not slept or returned to her hotel room, except for a shower and change of clothes. She'd been living on small cans of coffee, the caffeine fueling endless negotiating sessions.

Now, with chief U.S. negotiator Stuart Eizenstat, Under Secretary of State for Economic, Business and Argicultural Affairs, she was still fighting with other national representatives over basics. The U.S. wanted to stabilize levels of CO_2, methane and four other greenhouse gasses at 1990 levels by 2012. It wanted meaningful participation by developing countries. McGinty and Eizenstat pushed for "emissions trading," joint implementation and carbon "sinks," forest or ocean cover, to be figured in when national emissions caps were fixed.

As Rafe Pomerance, another U.S. negotiator, said, "The bigger the target, the more provisions you had to have in there to lower the costs."

The 15-member European Union, angered by what they regarded as a U.S. position filled with loopholes, demanded a 7.5 percent reduction of carbon dioxide, methane and nitrous oxide from 1990 levels by 2005, and 15 percent by 2010. They opposed cutting other greenhouse gasses and weren't about to ask developing countries to make cuts.

Japan wanted targets geared to per-person energy usage, not national averages. That way, energy-efficient countries like Japan would be rewarded.

A coalition of vociferous developing countries—refusing to make cuts until they were more developed—called for a gradual 35 percent reduction by industrialized nations in the three major greenhouse gasses by 2020. They wanted a special fund established to reimburse OPEC countries for lost income if the protocol came into force.

One Brazil minister said developed countries would adopt mandatory targets eventually if richer countries took them on first, for as long as a decade.

"We feel like there was a dinner and we were invited afterwards to have one cup of coffee, and then they asked us to pay the whole bill," he said.

The absurdist elements were heightened as the last hours ticked away. Observers downstairs were left to contemplate years of gamesmanship; the theatrical posturing that always marked early sessions. The way important ministers never arrived until late. The way agreements always seemed to be reached just in time to avoid utter failure, but without committing countries to action.

Even the setting underlined comic elements. Unlike international peace negotiations, usually conducted in quiet locations, climate business always seemed to be transacted in some gigantic hall filled with demonstrators and journalists, as if, a cynic might decide, the real goal was symbolism, not substance.

And then there was the tangled treaty language that had evolved over years, a tongue-tying collection of specialized terms that seemed to substitute for progress. Journalists covering proceedings had to master a special lexicon just to follow a five-minute discussion, peppered with references to "Annex One Countries," "Conferences of the Parties," "G-77 countries," protocols, declarations, mandates.

In some ways the process had evolved into an international holding pattern, but it was the only game in town.

Now, downstairs in the conference center, away from the key negotiations, mobs crowded around large TV screens showing live coverage of public negotiating sessions, swarmed toward staffers handing out press releases, and held demonstrations against opposing sides.

Abdullahi Majeed, chief meteorologist of the Republic of Maldives, found himself in one meeting room, once again stewing as he listened to an OPEC delegate, a Kuwaiti this time, talk about alleged benefits of greenhouse emissions, like the way more carbon dioxide would help crop growth and make growing seasons longer in northern areas.

Majeed, unable to take it anymore, said, "One hundred million environmental refugees? Some countries underwater? I don't think it right to talk about a few developed countries that might benefit."

The Kuwaiti fell silent. The diplomat to Majeed's left leaned over and whispered, "Thanks for shutting him up."

Upstairs, Katie McGinty exhaustedly moved between small rooms on the second and third floors, as she and Eizenstat held intensive separate sessions with the leadership of the EU and the Japanese.

"There was a lot of brinkmanship."

The Europeans balked at emissions trading.

"I was in this dark little room where night was day and day was night. The U.S. decided if we were going to face this obstinate wall with the EU, we would try to stitch together an agreement either the EU could sign onto or not . . . We started to talk to Australia, New Zealand, the Ukraine, Russia . . . When we had a united front, we went back to the Europeans," McGinty said.

Eizenstat then asked the EU to agree to cuts that were 1 percent higher than whatever the U.S. took on.

"He said that's the way they could sell it in America, on this competition business," said John Prescott, deputy British prime minister, who was in the room. "And the Japanese wanted one percent below the Americans."

The angry Europeans kept bringing up Chuck Hagel and the opposing U.S. senators downstairs. Kept throwing them in Eizenstat and McGinty's faces. To the Europeans it seemed the U.S. had sent two opposing delegations. While the official one negotiated, the one led by Hagel announced that any agreement McGinty and Eizenstat negotiated was meaningless unless the Senate approved.

"It disrupted negotiations because other countries were disinclined to make concessions if they felt they were making them for naught," McGinty would recall. "That an agreement's not worth the paper it's written on because some representatives of the U.S. are saying that they will give no credence to it . . . We'd get that all the time."

Resentment had certainly been swelling against the senators downstairs. At one point Hagel had found himself in a heated exchange with Conference Chairman Raoul Estrada of Argentina, a strong proponent of a treaty.

Estrada told Hagel that "the United States Senate was mucking around in international affairs it didn't understand," angrily recalled Hagel's aide, Deb Fiddulke.

Estrada countered, "I was steering the negotiations and they were trying to stop them . . . I'm afraid some members of the Senate were reflecting the positions of industry which were . . . supporting their campaigns.

"I had an opportunity to show a different point of view, from the reality of the world, which is a little bit larger than the U.S. itself . . . I've had an opportunity to live in the U.S. and you can see how the air conditioners are always working there. How the heating is working. How the lighting in houses is on twenty-four hours a day, seven days a week. This is not efficiency . . . I was trying to explain *that* to the senators . . . to show them their mistake."

As midnight passed, Estrada went upstairs to the rooms where Europeans and Americans were arguing. He walked into chaos. Deputy Prime Minister John Prescott was waving his arms so hard at Stu Eizenstat that his chair broke. The Japanese were arguing among themselves. Their foreign-ministry representative "had to negotiate with their industry and their environment ministry at the same time he negotiated with other countries."

Finally, though, "The numbers were worked out in that room," Estrada said.

The group agreed to 8 percent cuts for the EU, 7 percent for the U.S., 6 percent for Japan. John Prescott, convinced by Al Gore that emissions trading was a legitimate means of cutting emissions, and not a loophole to avoid it, helped push the European Union to accept it.

As for other big questions, like who would enforce a treaty, they would be negotiated later.

As the last hours ticked down, the draft proposal, distributed for approval of the overall body, contained a U.S.-backed provision to allow developing countries to voluntarily take on targets later. The U.S. team hoped the Senate might regard this as "meaningful participation."

Now the final big session began, a line-by-line examination of 28 articles and two treaty annexes. With so little time left, any delaying tactics would kill the treaty.

Sure enough, opposition to the stitched-together agreement began as soon as emissions trading came up.

But now, in a bizarre end to the meeting, Raoul Estrada decided to push through the agreement on his own. Each time opposition rose to a point, he'd either announce a brief adjournment and work out a compromise, or bang his gavel and simply announce the article deleted or passed.

In the end, the bulk of the U.S. requests were reflected in the protocol. Reforestation and sinks would count toward reductions. Targets would be reached between 2008 and 2012. Reductions would be mandatory for all six major greenhouse gasses. Emissions trading would be permitted, and industrialized countries could get credit for building energy-efficient plants in developing countries.

But the reference to voluntary emission cuts by developing countries was deleted.

Nevertheless, from now on, the Kyoto Protocol, small step that it was, would make it a bit more difficult for corporations and governments to ignore global warming. The pressure to deal with it had just been turned up.

Governments that had been hiding behind U.S. opposition would now have to ratify the treaty or look like hypocrites. Corporations would start proposing alternate solutions to keep the protocol from becoming stronger, something that could easily happen if weather kept getting worse, or the next IPCC report concluded that new evidence linked human activity to deteriorating climate.

Flying home, Katie McGinty knew she faced the next round of climate battles. It was one thing to agree to a treaty, quite another to get the Senate to ratify it. In 1998, a furious Chuck Hagel intended to block the treaty, and opponents in Congress planned to initiate legislation to limit *any* actions the Clinton administration might take, even indirect ones—in bills and appropriations—to work toward energy cuts.

Hagel and Senator Robert Byrd would call upon President Clinton not to sign the protocol, and although the U.S. would in fact sign it, the Clinton White House would never submit it to the Senate for ratification. They would put it aside and hope for a change in Congress or in the position of developing countries—before offering the Senate a chance to confirm or kill it. Otherwise submission would have been suicidal. For the Kyoto Protocol, as for the doomed League of Nations after World War I, there wasn't a chance in hell the Senate would ratify the agreement the White House had made.

Massachusetts ... The Arctic ...
Honduras ... New York

*T*WO HOURS BEFORE MAKING his groundbreaking discovery in climate science, which would turn out to be the biggest in the 2000 IPCC report, Michael Mann warned himself to drive slower, even though he was already moving at a crawl. The 32-year-old University of Massachusetts postdoctoral research fellow knew it was safer to stay home in the storm but he was too excited to do that, too eager to finish the project on which he'd been working for the last three years.

It was January 1998, the hottest January on record, and in northern Massachusetts, with temperatures higher than usual, a winter storm that normally would have covered Amherst with snow was coating it with ice instead. During his one-and-a-half-mile drive to campus, ice turned Mann's windshield wipers scratchy and covered the fine old Victorian homes he passed with a thick glaze that shimmered on turrets and gables. It made roads slick and snapped off oaks, maples and power lines. The traffic light at Commonwealth Avenue and North Pleasant Street dangled with stalactites of ice.

His hat was pulled down, his overcoat buttoned even in the car. Mann switched on National Public Radio to hear that in Quebec, Canadian troops had been called out to deliver food to isolated communities and make sure elderly people didn't freeze without power. Insurance losses were already skyrocketing. They would exceed $1.2 billion across the northeast U.S. and Canada.

By the end of the day, Mann would be able to say lots about why weather extremes were rising across the planet, and due to his work, so would the IPCC. Reaching the University of Massachusetts campus, the former Yale grad student gingerly parked and made his way along slippery paths to the office he shared with two other postdoctoral research fellows, neither of whom had shown up in the storm. In the toasty room he put on a Van Morrison tape and eagerly pulled a seat to his desk, piled with stacks of reports plotting temperatures over the past 600 years, reconstructed from ice cores and tree rings around the world.

The project idea had been Mann's and Dr. Raymond Bradley's, the professor who had brought Mann to UMASS after they met in 1995. Ironically, the study they had initiated back then had nothing to do with climate change.

"I was curious to understand natural variability in climate, not global warming," Mann said. "The words 'global warming' did not appear in the proposal we sent the Department of Energy. It was not even a topic under consideration."

What *was* under consideration was the use of tree rings and ice cores—called proxy records—to get an idea of how the earth's climate varied naturally long before the concept of a man-made "greenhouse effect" was ever possible.

The ice cores were yardlong, four-inch-wide samples drilled from glaciers on all continents. Air bubbles frozen inside contained samples of ancient climate.

The tree rings would be analyzed at the University of Arizona tree ring laboratory, where scholars could glean from their width, density and composition the makeup of earth's atmosphere as far back as when the black death was ravaging medieval Europe.

Mann and Bradley, in short, would work from real evidence. They would not be making computer simulations.

Bradley was the expert at collecting the proxy records and forming them into databases. Mann, the statistical climatologist, would combine the data via computer and see if the ice and tree records matched and whether they would confirm or disprove the re-created ancient climate records provided to researchers by computer models, the main means so far, and the hotly disputed one, of estimating temperatures before instruments existed to record them.

After the first year on the project, two years ago, Mann had excitedly found that the data *was* matching up, *was* confirming computer re-creations.

By that point Mann had graphed temperatures over the last 600 years. His graph "resembled a stock market line, wiggling around for 500 years, and suddenly you get to the year 1900, and it's no longer wiggling randomly. It starts going up. It keeps rising. For the twentieth century, the rise took us out of the range of all previous wiggles."

Which confirmed that the 1900s were the hottest century on record, but it still did not mean that humans were responsible, or that the high twentieth-century temperatures were outside the range of natural variation. The real question was whether so much heat—even unprecedented over a century—was inside or outside the naturally possible range.

At that point, the project evolved to a next stage. Could Mann reconstruct global *patterns* of climate from the records? If so, they would provide a test for modelers, an error bar, a way of measuring whether their reconstrucions were in the naturally possible range.

So by early 1998 Mann had produced from the proxy records a "range of natural variation" superimposed over the temperature graph between the years 1400 and 1900, a shaded area on both sides of the temperature line looking like a floodplain on either side of a river.

If temperatures stayed *inside* the shady area, they were within the range of natural variation.

But if they jutted *outside* the shady area, they surpassed natural variation, and therein lay Mann's clever test. For the centuries preceding the industrial revolution, every influence on climate had been natural—solar energy, earth's orbit, clouds, volcanoes—*everything*. Therefore it should be impossible for temperatures to move outside the natural range. If they did in the graph—if the temperature line meandered out of the shady area, it would mean that the scientists who had estimated past temperatures had been wrong. They would have to start all over again.

But Mann's reconstructed temperatures between 1400 and 1900 had fallen *inside* the range of natural variation, which made it much more likely that computer models were right.

Mann's project, so far, had confirmed IPCC findings, although he would have been equally gratified if it hadn't. He just wanted to know the truth.

Now he and Bradley were down to the last big step.

"We decided to look at the 1990s and say something about how unusual they might be in the context of a reconstruction."

So today Mann would extend his natural variability test to temperatures recorded after 1900, long after the effects of the industrial age had begun to be felt in the atmosphere. Today he would see if the computer model reconstructions, proxy data and records for the twentieth century fell within the natural range too. If they did, then the global warming that the earth was experiencing in the 1990s was very possibly natural in origin and *not* humanly caused.

But before he ran the program, he flipped on his computer and e-mailed a colleague in Canada, "Do you have any updates discussing your ice core work?"

After all, until the very last minute he wanted to collect as much raw data as he could, to cross-check all aspects of his work.

The frantic answer came back. "I have been trapped away from work the last few days and cut off by this mega–ice storm that hit Ontario, Quebec and upper New York. I've never been in a disaster situation before and I can't recommend it . . . Fifty percent of the deciduous trees in eastern Ontario have lost over half their branches and might die. No shit. Weather permitting, I'll be in next week."

Mann thought it sounded pretty bad up there. But there was no new information.

He sat back and began tapping at his keyboard, creating the newest temperature/natural variation graph, while Van Morrison sang on the tape, ice machine-gunned against the window, and the wind soughed outside.

Up swam the color graph on Mann's screen. He sat there, staring at it, dumbfounded.

"Holy shit," Mann said.

On the graph, the black line showing temperatures throughout the 1900s zigzagged steadily upward throughout the early part of the century, then turned steeply and rocketed toward the top of the graph. On both sides of the line, for a while, he saw blue dashes, the limits of natural variability. The area between the dashes was shaded light blue. But at the end of the twentieth century, a bright red trace line, late-twentieth-century warming, shot out of the blue zone, *out* of the range of natural variability.

Which meant that outside of some fluky un-thought-of answer, *humans had caused the warming.*

"For the first time, we could place the modern record in the context of the past," Mann said.

Mann began firing off e-mails to everyone who had worked on the project.

"We'd stumbled onto something special here."

Outside his window, as he typed, the ice storm kept up a steady assault on the school, town, state and region. Mann looked up to see his wall calendar beside the window, showing for January, a typical winter scene, a photo that had been taken during a "normal" New England year.

The calendar showed snow on the ground. Evergreens covered with thick, white, heavy snow.

Not a rain of ice.

———

February 1998 turned out to be the hottest February on record. March the hottest March. April the hottest April.

The 1998 El Niño was turning into the worst on record. Major floods would strike Korea, eastern India, Bangladesh and New Zealand that year. Heavy rains in China would kill 3,000 and leave 230,000 homeless when rivers overflowed. Heat waves and air pollution alerts would sicken thousands in Egypt, the Mediterranean area and southern Europe.

Brush fires set to clear farmland in Indonesia would burn out of control when the usual monsoon rains failed to arrive and put them out. The flames would consume forests, drive refugees to other islands, destroy half the coffee, sugar and cocoa harvests, and create respiratory problems for people occupying an area thousands of miles across.

"The exceptional nature of the 1997–1998 El Niño suggests that it was aided and abetted by . . . the warming of the earth . . . from the accumulation of carbon dioxide and other greenhouse gasses," a report called "Consequences," published by NASA, the NSF and the DOE, would conclude.

The economic cost of weather disasters that year would top $89 billion—a record in itself, according to Munich Re, the world's largest reinsurer. The previous record loss from natural disasters, $60 billion, had occurred in 1996.

But financial losses were only part of the problems from rising heat.

By 1998, an army of researchers were linking losses to earth's biodiversity—its mix of species—to rising temperatures. Like the planet's human population, plants and animals had evolved to survive natural swings of climate. Unlike humans, plants and animals lacked technology to protect themselves if climate changed too fast.

Adélie penguin populations were dropping in Antarctica, as sea ice melted. Polar bear births were down 10 percent in Manitoba because it took ice longer to freeze on Hudson Bay each fall—keeping the bears from walking to their feeding grounds. Southern California zoo plankton, ocean microorganisms that feed fish and sea birds, had dropped 70 percent since the mid-1970s, and, not surprisingly, so had many fish and sea bird species. Warbler populations were dying out in Minnesota and Alaska. Amphibian and bird counts plummeted in the cloud forests of Costa Rica.

In 1998, also, coral reefs began dying as ocean temperatures heated up. Ninety percent of all corals in the central Indian Ocean died from bleaching, and mass bleachings began killing off reefs in all ten "reef provinces" around the world.

In places like the Maldives, famed for their reefs, divers who would have normally seen a riot of color and movement underwater—5,000 species of shell, 200 species of coral . . . loggerhead turtles and leatherbacks . . . surgeon fish . . . parrotfish . . . butterfly fish—saw that the color was gone, and at the moment so were the fish.

Coral reefs are a base of ocean life. Although they look as dead as inanimate rock to casual divers, they are actually composed of skeletal deposits from billions of tiny creatures that live in the skin of the coral—nicknamed "zooks"—that give the reefs their bright color.

Coral reefs allow fish to breed. They attract tourists to strengthen economies. They provide the U.S. National Cancer Institute with chemicals to test against cancers, and constitute a base that keeps thousands of people in the developing world from fleeing if their environments collapsed.

"Coral has adapted to natural ocean temperature swings over millions of years. But the rate of change in 1998 was horribly above the threshold that coral could survive," said Susan Clark, a British researcher in the Maldives that year, hoping that the ocean temperatures might cool a little and allow the reefs to begin to come back.

"There are no clear records of mass bleaching events to this extent in recent history," wrote Mark Spalding of the World Conservation Monitoring Centre in England. Spalding predicted that high temperatures that drive mass bleaching might begin occurring every year by 2030, if IPCC estimates of coming global warming are right.

———

May 1998 became the hottest May ever, and June the hottest June. As earth burned, and evidence accumulated that humans were responsible, at 6:00 A.M. one morning in July that year, the hottest July on record, one *more* piece of the puzzle edged toward falling into place when a cheery New Hampshire–based geophysicist, outdoorsman and father of three named Don Perovich awoke to his ringing alarm in a small, comfortable ship's cabin featuring a real bed, sink, couch, desk and thick red parka hanging within reach on the wall.

He pulled down the parka and peered outside the porthole it had covered to keep the sun from waking him. He knew that ice would stretch in all directions out there, but his immediate question was, would it be in the same place it had been yesterday? Or had the whole ice camp outside—including a 60-foot-high antenna, tents, a hut for storing ice drills, sleds and optical instruments, a Quonset hut garage and three ice runways—drifted away overnight?

"One day in March we looked out, and the camp had moved a quarter mile off."

But the camp was there.

Satisfied that he faced no immediate problems, Perovich got ready to make climate science more exact.

He was on the Canadian icebreaker *Des Groseilliers,* which had been intentionally locked into sea ice 320 miles north of Deadhorse, Alaska, nine months earlier, as part of the largest, most complex project ever supported in the Arctic by the National Science Foundation. The whole idea of the $19.5 million SHEBA project (Surface Heat Budget of the Arctic Ocean) was to let the ship drift with the ice in a unique strategy to study the atmosphere. It would travel 1,800 miles by the end of the year.

"SHEBA was about providing information to make climate models more accurate . . . It was not about proving global warming," said Perovich, chief SHEBA scientist, on loan that year from the U.S. Army's Cold Regions Research and Engineering Laboratory in Hanover, where he usually lived.

On ice station SHEBA, 25–30 scientists were minutely monitoring the "SHEBA column," an imaginary cylinder stretching far above and below the ship, from the top of the atmosphere, miles up, into the lower atmosphere, through the ship and ice and finally down into 600 feet of ocean.

"We wanted a vertical profile of the atmosphere: temperatures, wind speeds, humidity." Everything.

The basic information would go to people like Jim Hansen, and help fine-tune climate models—address the problems that greenhouse skeptics had with them.

SHEBA "cloud teams" were examining clouds in the column, to determine whether they helped cool the earth by limiting the amount of sunlight reaching the surface, or warm it by acting as a blanket to keep heat from escaping off the ice. Greenhouse-skeptic scientists said clouds *cooled* the earth. Believers thought the opposite.

Perhaps SHEBA could help resolve the conflict.

The "ocean team" monitored temperatures and currents in the sea, and how they changed during the year. That way, links between ocean and atmosphere could be more accurately programmed into models too.

The "ice team," Perovich's, studied ice around the ship, to see how much it grew in winter, how much it melted in summer. They also carefully measured the "albedo" effect, an important climate feedback. Arguments about albedo—the amount of the sun's energy reflected or absorbed by different kinds of surfaces—peppered climate debates.

"If you have a surface that's all black, like the ocean from space, it'll *absorb* all sunlight and make the surface warmer," said Perovich. "Its albedo will be zero. If you have the brightest incredibly white surface like bright snow, that *reflects* sunlight, its albedo will be one . . . We're very concerned about how albedo changes over time . . . When snow warms and melts, its albedo gets smaller and it melts quicker. Being able to quantify that is important. How sensitive is the whole system if the melting starts a week earlier? Will it make a big difference or not?"

Now, cheerily dressing to a "shhhhh" sound (blocks of ice rubbing against each other outside) and a sharper gunshot sound (ice cracking), SHEBA's chief scientist bore a striking resemblance to New York mayor Rudolph Giuliani—in his high forehead, black thinning hair and pale roundish face, except in Perovich's case the prosecutor's sternness was softened into an almost Moonie-like friendliness.

Today's day-in-the-life of an "ice person" would pit him, as he worked on climate, against fissures in the ice pack, possible attacks by polar bears and the ever present summer danger of falling into the near-freezing sea, which at this location was 11,000 feet deep.

"Almost all the scientists fell in at one time or another."

He'd been working for 20 years in the Arctic by 1998, flying into ice camps in Alaska and Canada, living in tents, spending nights in sleeping bags, storing food outside.

"The old saying in the Arctic is, when the ice cream melts outside, it's time to go home."

Working with ice had always "sounded neat, the idea of being able to walk around on frozen ocean." And as the ship slowly drifted clockwise around the North Pole, Perovich's boyish enthusiasm extended to softball games on the ice, lectures on the ice, even meals on the ice.

"The first rule of the Arctic is, never get the chef mad."

After gobbling down the grilled-cheese breakfast ("Delicious!") he went back to his cabin and changed into lightweight long underwear, a flannel shirt, army surplus wool pants and hip boots. His red Gore-Tex jacket was new, as was the black hat and Gore-Tex Windstopper gloves.

"In the summer you're in a giant icebox out there, so no matter how much the sun shines, the ice determines the temperature. It always hovers around freezing."

He also took a blue knapsack filled with emergency rations (M&M's),

Global Positioning System equipment (if it got stormy, you couldn't see the ship), a bottle of water and an orange lifejacket.

Then he went by the bridge to pick up a portable radio and shotgun: the radio for communicating with the ship if he needed supplies or an emergency arose, the shotgun to scare off, or in a worst-case scenario shoot, a polar bear.

At 8:45 the five-person ice team left the ship and opened a steel door at the bottom of the gangplank, there to keep out bears. Perovich saw a panorama of ice and open water, the water black, the ice white. Meltponds, small pools atop thicker sea ice, were aqua, turquoise and light blue. The summer sky was a uniform gray, low and overcast, and surface contours in the distance, generally flat, occasionally pushed themselves into hummocks or ridges.

The ice team—three men and two women—split up and headed out on foot or snowmobile toward measuring points they visited daily: "Seattle," "Pittsburgh," "Quebec."

"The rule is, nobody goes out alone."

Perovich and two female scientists hiked a quarter mile to a storage hut, and picked up a banana sled, a six-foot-long enclosable carrier on wooden runners. They loaded it with a generator, voltage regulator and ski poles with tape marks on them to monitor pond depth. They man-hauled the sled to "snow lines" where, at periodic points, stakes stuck out of the ground—as landmarks—and a way to track surface melting.

To measure changes in ice mass, Perovich kneeled down at holes that had been drilled months before, when the ice was thicker. He'd grasp a handle from which a wire dropped down through the ice to the sea. Attached to the other end of the wire were crossbars. By measuring how much wire he could pull from the hole before the crossbar hit ice, he learned the thickness of the ice.

"In July, the ice was melting faster than we thought it would."

By noon, the team was doing "edge measurements." One person would step off the surface ice into a melt pond, onto a weakening ice shelf, wearing crampons to cut down on slipping. He or she would use a marked ski pole to determine how much the ice had melted there since the previous day.

"The ponds were melting all the way through and becoming black holes," Perovich said.

At one, lunchtime normally, an unexpected clearing moved the clouds off, and Perovich radioed the bridge to order a small helicopter made ready for an aerial survey before the clouds massed in again. Soon he and two other ice people were 6,000 feet up, flying a 30-mile square around the ship, snapping photos of ice and recording remote temperature measurements of the surface.

"Even though we're objective scientists, we were rooting for the ice not to melt any more. You wouldn't expect a group of people in the Arctic to say, 'I sure hope it gets colder.' But we sat around doing a lot of that."

By two Perovich was in the mess hall, eating.

"Mmmmmm. Delicious tomato soup!"

By afternoon, he was back outside on the "albedo line," a 600-foot path marked by stakes every 150 feet, which had been part of an area that had drifted away from the ship since it was first established.

"It went from being less than a mile away to three miles. In the beginning we got there by walking, then by rowboat and kayak, and by the end, by helicopter."

As they worked, taking measurements every seven feet, the radio informed them that a research plane from another project would shortly be flying overhead. Within minutes, through the vast Arctic silence, Perovich detected a growling sound in the sky, and a big C-130 airplane roared toward them, flying low. It was based in Fairbanks, Alaska, making a series of crisscross flights, also measuring ice thickness and albedo with remote sensing equipment.

The big airplane flew into the SHEBA column. All over the world that year, dozens of scientific projects were examining the same climate questions in different ways, cross-checking each other, recording results that would sharpen climate models and that would go to Tom Karl and other scientists working on the 2000 IPCC report.

The plane flew back to Fairbanks.

An hour later, Perovich's albedo work was interrupted again by five long blasts of the ship's horn. Bear alert!

"This one turned out to be a mom and cub. Usually in an alert, someone goes out on a snowmobile to escort the bear away, making sure, through the noise, that it doesn't get close to the ship. One time a mom and cub started flossing their teeth on a cable. They looked like a couple of strong cats playing with a string."

The bears moved off without bothering anyone.

At five, 11 hours after his day had started, another blast from the ship signaled suppertime, after which, this being Monday, the scientists held their informal weekly meeting in the ship's bar/lounge to share results learned over the last seven days. By mid-July, it was very clear to them that the ice was melting much faster than anyone had anticipated.

The atmosphere team started today's discussion.

"Look at this! The results of the past week show unusually warm air up several hundred feet."

"Where'd *that* come from?" asked someone in the ocean team.

"Must be global warming," quipped a member of the ice team.

"Nah, this is weather we're seeing, not climate. Maybe a storm coming from the Bering Strait."

"Yeah, well, I wonder what we're all going to do when the ice cap melts?" someone else said.

Perovich joked, "I just hope it doesn't melt until my kids get out of college."

Under the laughter there was an edge.

The final SHEBA report, titled, "Year on Ice Gives Climate Insights," would say that Perovich and other SHEBA people "witnessed . . . thin ice at the start and even thinner ice at the end," of the project. "Also, the extent of open water during the summer of 1998 in the Beaufort and Chukchi Seas was the greatest of the past two decades."

When it came to the disputed role of clouds on global warming, the report would say, "One of our significant findings was that the Arctic summer clouds warm the surface and enhance ice melting . . . the blanket effect dominates."

Which meant, if the phenomenon was widespread, that more serious estimates of future global warming from greenhouse gasses were more likely to be accurate than less consequential ones.

When it came to ocean temperatures, "measurements made at this time found that the upper ocean was warmer and fresher than 20 years earlier."

Overall, the report concluded that, for the SHEBA scientists, despite their great knowledge of the global warming debate, the thinning ice had been a "surprise."

When it came to surprises in 1998, Bill O'Keefe and the American Petroleum Institute got a big one on October 5, during the first month of that year that did not break a monthly record for heat.

"Climate Change," read the big ad in the *New York Times* and *Washington Post* that O'Keefe read that morning. "Waiting only makes things worse!"

The argument was old, but the signatories on the ad were new. It wasn't some environmental group this time, but BP America, Toyota, Lockheed, Maytag, 3M, Weyerhaeuser.

"The longer we wait to address climate change, the more costly the process will be—to both the environment and the economy," the ad said. "That's why some in the private sector want to act early and want government to provide credit for early action."

In 1998, for the first time, a united group of major corporations were acknowledging that climate was changing, humans had caused it and action was needed to slow it. And since climate political initiative always seemed to follow public or corporate pressure, the new action was groundbreaking.

Behind it was Eileen Claussen, the ex–treaty negotiator, who had been busy since quitting government. Earlier that year she'd been approached by the Philadelphia-based Pew Charitable Trust and asked what she thought should be done about climate change.

Her answer was that corporations needed to be involved in the process, that many were ready to acknowledge a problem, and that they had many legitimate concerns which, if adequately addressed, could push some solutions forward.

She suggested that Pew establish a Center on Climate Change to accomplish these things, and, "They said, 'How'd you like to run it?' " Claussen said.

Pew funded the center with a $5 million annual grant, and by fall of 1998, Shell Oil had left the GCC and joined the center. DuPont joined too.

Individually, Pew companies announced they would or had already initiated policies to cut down on greenhouse gas emissions and energy consumption. They promised to invest in efficient energy technologies.

Their programs were as controversial as American Electric Power's plan to increase nuclear power production; as simple as Pratt & Whitney's $3,600 investment in decals asking employees to turn off lights at night (which saved $250,000 a year); as innovative as Pacificorp, BP and AEP's purchase of Bolivian rainforest to preserve it; and as complicated as BP's internal system of emissions trading between different divisions around the world.

"The best approach is for business to lead, not wait for government mandates," said Dennis Reilly, DuPont's executive vice president and chief operating officer, adding that by 2010 the company hoped to obtain 10 percent of its global energy from renewable resources like wind or solar power.

Pew companies also lobbied for bills that would reward energy reductions. Such laws, they argued, "would provide companies with more certainty and incentives to act."

For instance, Pew companies wanted credit for cuts they made now in case the U.S. committed to making mandatory ones later. Otherwise, the companies would have to make even *more* cuts later, so why do anything now?

Enron joined Pew. So did Air Products and Chemicals, Inc. PG&E, Baxter International, ABB. A dozen Fortune 500 companies in all (and 28 by 2001) as part of the slow creep toward solutions.

Partially as a result of corporate interest, three senators—Democrat Joe Lieberman of Connecticut, Republican John Chafee of Rhode Island and Republican Connie Mack III of Florida—introduced legislation in Congress to give credit to corporations for energy reductions.

Under the bill, the government would give ton-for-ton credit to any of the 150 U.S. companies that could document reductions in greenhouse gasses under various voluntary federal programs of which they were already a part.

With U.S. emissions projected to rise 30 percent by 2008, under the IPCC's "business as usual" scenario, the bill didn't pass, but even conservative Republicans in Congress began paying at least lip service to market-based solutions.

"There's been an undercurrent of honest exploration," one GOP staffer told *BusinessWeek.* "Industry knows it can't say, 'hell, no,' forever."

Meanwhile, in 1998, across the Atlantic, the European Union reached an agreement with carmakers to reduce average fuel use per new car by 25 percent in Europe.

It was becoming harder for U.S. carmakers to say such goals were impossible. It was becoming harder for *anyone* to argue that climate change was not going on.

———

Roger Omar Hernandez, who was about to experience the most terrifying weeks of his life, was a compact, dark-haired field supervisor for the Chiquita banana company in Honduras. He carried himself with the crisp, confident movements of a well-conditioned laborer, and was a soft-spoken family man with a wife and three children, a religious man, and a hard worker respected by his bosses and the 28 field hands who worked for him on "Ceibita," one of 19 company banana plantations in the Sula Valley, one of the finest banana-growing areas of the world.

Thirty-eight years old, Roger had never heard of environmental security expert Gary Vest at the Pentagon and his prediction that extreme weather events could soon destabilize huge chunks of the third world. On the rainy morning of October 28, 1998, even if he had heard of Vest, he would not have suspected Vest was talking about him.

Ceibita occupied a low, hot, green valley, flanked by high mountains visible to the north and south. The plantation lay between the Ulùa and Chameleon Rivers, both close enough so Roger could walk to them, both silty brown, both meandering to the Caribbean along the same general route that Chiquita's trucks took daily, transporting bananas Roger's crews picked—part of the 20 million boxes a year the company sent from Honduras to the U.S.

Housewives in Harris Teeter supermarkets in Raleigh, North Carolina, took Roger's bananas home to put on their cornflakes. A&P customers in Philadelphia and New York sliced Roger's fruit onto their hot-fudge sundaes, their flan and cream pies, and into their big bowls of fruit salads on hot summer days.

The valley was rainy in October, dry in spring—traversed by a narrow dirt road dotted with barefoot children, workers on one-speed bicycles, trucks filled with sugar cane or bananas and late-model cars or four-wheel drives owned by Chiquita officials or by the company.

Roger lived in company housing on the farm, in a neatly laid-out town, its streets clean, its stop signs new, its homes two stories high, of wood faded by the tropical sun. The elementary school 100 feet from Roger's house was attended by children who greeted guests, company officials, doctors or, within weeks, horrified reporters from around the world, with a cheery coordinated "GOOD MORNING!"

In the fields, banana plants sprouted in orderly rows, bouquet-shaped, their stems thick, their fronds wide, their bunches covered with plastic bags to make them ripen more quickly from a man-made green-house effect, as Roger's men tied color-coded ribbons around each bunch denoting the maturity stage of the fruit inside.

That October Wednesday began no differently than other ones. Rain or shine, bananas need tending. Working carefully, so as not to slip in the mud and cut themselves with their machetes, the men performed their usual range of activities. Chiquita farms were divided into sections and field bosses were in charge of every aspect of care in the sections they oversaw. As Roger gave orders, his men tied guy wires to "mother plants" to prevent blowdowns from wind damage. They cut off leaves that could scar fruit. They "dehanded" plants, slicing off sections that underproduced bananas. They monitored the growth of "fingers," indi-vidual bananas, and that of the clusters on which they grew.

"It had been raining for a few days," Roger would later recall. "Then the rivers started overflowing, so we alternated between sand-bagging the river, and taking care of the bananas. In October, if it rains, we always watch the rivers."

During the farming periods, harvesters worked in two-person teams. One man would stand beneath a bunch and hoist the bottom over his shoulder. The second man would cut the stem so the bunch eased onto his partner's shoulder, and the bananas would not be bruised.

As it kept raining, in the nearby city of La Lima, Chiquita banana officials were on the phone to the U.S. National Hurricane Center in Miami, where an automated message gave news of Hurricane Mitch in the Caribbean, which several days earlier had been off the coast of Venezuela.

Arnaldo Palma, head of all Chiquita operations in Honduras, had been checking on the storm twice a day.

He was a quick, intelligent man with decades of experience with bananas. "There is a hurricane period that runs from August through September and any time you have a tropical depression during that period you follow up on it. Tropical depressions can turn into hurricanes. Wind over 30 miles an hour starts damaging us."

Winds in Mitch had been blowing over 30 miles an hour since October 24, when the storm had become a hurricane, 255 miles southwest of Kingston, Jamaica. They had intensified until by Monday, the 26th, two days ago, Mitch had become a category 5 storm, its winds exceeding 155 miles an hour.

At first, the Miami recordings had assured Chiquita that Mitch would not head in their direction, but the recordings had *not* included the proviso that slow-moving tropical storms are more difficult to predict.

"The message told us the hurricane would take a north/northwest track and hit somewhere between Cuba and Mexico," Palma said. "We tracked it with pins on a map."

The forecast had been so reassuring that Roger's boss, Ceibita's manager, Jorge Fajardo, had flown to the U.S. with other Chiquita executives to visit customers. But on Wednesday, as Roger's men picked bananas, Fajardo was in a New Jersey hotel room, glued to the Weather Channel, realizing the same thing that Arnaldo Palma was seeing in La Lima.

The hurricane was heading for Honduras.

During those last minutes before company radio began broadcasting evacuation warnings to the farms, banana work proceeded as usual. In the packing plants, conveyer belts moved banana bunches into tubs filled with fungicide solution. Rows of chatting women in yellow aprons, including Roger's wife, gently arranged the fruit to avoid bruising it, in cardboard boxes lined with polyethylene bags.

Hurricane Mitch had reached the Bay Islands off the coast of Honduras. It had weakened from a category 5, but rain continued to fall.

"On the company radio, we heard there will be a flood and a hurricane," Roger said.

He told the workers to go home. Everyone left work on all the farms. Some families began evacuating.

By the time Roger reached his house, "there were two inches of water in the town."

Still, "Even though the radio said to evacuate, I was thinking, there's no need. People told me that it never floods here. There had been a storm in 1974, but the flooding was very little then."

For Roger, it was not just a question of inconvenience. He knew that if you evacuated, you left behind your belongings for thieves. Also, evacuees had to walk miles through rain to get anywhere.

"So I carried my furniture upstairs: stove, refrigerator, dining room table and chairs, food and groceries, all of it. The whole family carried things."

After all, he figured the second story of their home was a good 12 safe feet above the ground.

"But when I looked out it was neck deep down there. I thought, maybe I should have left. Now it was too late."

At least the family had electricity, but the radio was at the office, so there was no way for them to get news.

At nine, the lights went off. It kept raining.

"We started to get afraid."

Roger got little sleep that night. In the morning, looking out, the family was shocked to see their whole ground floor underwater.

"Water began rising into the second floor, fast."

Muddy water rose halfway up the portable stove and the refrigerator. The family had no choice but to leave their precious appliances to ruin as they retreated up to a small cramped attic, almost a crawl space, above the second floor.

"We prayed all day."

If Roger had had a radio, he would have learned that the storm, downgraded but dumping up to 35 inches of rain, was moving slowly through the country, following the mountains, heading for his home. Rain washed down the slopes on both sides of the Sula Valley, turning the rivers into torrents. In La Lima, Chiquita banana headquarters was underwater. Chief company physician Oscar Javier Perez Lazo, in charge of medical relief for the farms, horrified at the "permanent rain," was trapped and unable to send help. He was reduced to praying on his knees in his home.

"I prayed to the Mother and Holy Spirit," Dr. Lazo said. "I glorified the Lord first and thanked him for everything. Your will, will be done, I said, but please protect the community, the whole population of Honduras. Please calm the storm."

Arnaldo Palma, general manager, could only make phone calls and try to arrange for help—food, medicines and helicopters—when the rain ended. He could only sit helplessly while waters rose so high upriver that government officials ordered dams opened to alleviate pressure on them. The runoff raged toward Ceibita farm.

Roger Omar Hernandez, lacking a radio, had no idea what was coming toward him. By Friday afternoon, with electricity gone, candles gone, food gone and water three-quarters up the second floor of his home, he was trapped with his wife and three children in their attic. Then, as the dam runoff hit, they started feeling the house swaying.

"It was moving in the current. We heard dead cows hitting against the walls of the school."

Roger's wife began screaming. She couldn't take it anymore. She tried to hurl herself into the flood, to stop the terror. Thirty yards away, in the attic of the school, 100 people crammed into a crawl space could hear Roger and his wife fighting as he managed to keep her from jumping.

"I had some ropes. At midnight the waters were still rising. We all agreed at least we would die together. I tied everyone around the waist, to a beam. All night, until five the next morning, we stayed there with ropes around us."

For years, Roger would cry if he talked of it, how his wife had fainted from fear, how his family had called to the people in the school for help, screaming, "The house! It is collapsing!" How after daybreak, when they were up on the roof, and the house was swaying like crazy, he saw people on the roof of the school launch a small, wooden, makeshift raft toward him, as six men swam beside the raft, pushing it through whirlpools, past dead chickens and pigs whirling by, and floating garbage and snakes.

"They brought us, one by one, to the school. It took four trips."

It kept raining.

For the next 14 days Roger and his family lived in the cramped attic of the school, their wan light coming from candles, flashlights or little holes punched in the tin roof. They slept across beams. They shared, at first, chicken, frijoles and rice that people had brought with them, and when that ran out, men swam with the raft to coconuts, which they picked from the tops of submerged trees.

The people crammed into the attic went to the bathroom in garbage bags. They watched dead animals floating by all day. They had no way of communicating to anyone outside the farm, but transistor radio broadcasts told them that thousands were dead in the country, that the storm had stalled over Honduras.

Arguments broke out over whether the water was rising or not. People fell ill with diarrhea.

Once, a helicopter appeared overhead, and the people inside threw out plastic bags filled with food. But many of the bags were swept away by the current.

Mitch turned out to be one of the deadliest tropical Atlantic hurricanes in history. It killed 9,000 people in Central America. It destroyed 80 percent of all banana plantations in the regions, according to Swiss Re. Chiquita's losses reached $50–$60 million, some of which insurance would repay, after deductibles were met. But nobody would repay Chiquita for lost business.

Roger said, "I lost everything. But I am very grateful to God, because he gave me my life."

At Ceibita, within a week, food and medicine arrived by a tractor-pulled relief convoy. Chiquita began considering closing down all operations in Honduras, but ultimately would decide to stay, although some farms would still be out of operation in 2000—a lesson for U.S. corporations who own businesses in third world countries and will depend on infrastructure there during the warming century ahead.

Roger would remain a supervisor at Ceibita.

His wife would leave him, frightened that more storms might come. She would travel north illegally, to the U.S., with thousands of other environmental refugees fleeing Mitch's aftermath, crossing into Texas, Arizona, California.

Roger's wife would become part of the scenario that Gary Vest had envisioned, in Washington, at the Pentagon.

"My wife is a wetback," Roger would say at the turn of the century, still missing her, tears in his eyes.

In December of 1998, the month when a more lethal tropical visitor to the U.S. made its presence known, its discoverer, a petite, dark, attractive

Sri Lankan scientist named Varuni Kulasekera kissed her rock-and-roll musician husband goodbye, left their Brooklyn apartment in a snow-storm, and drove their Nissan Altima onto the Prospect Expressway and into Manhattan. Although it was a weekend she had been summoned urgently to the Upper East Side by officials of the New York City Health Department.

Varuni was fascinated by flying insects. She'd been that way since age six, when, reading a fairy tale book on the open-air verandah of her parents' home in Sri Lanka, she'd spotted an atlas moth, a real beauty, with a ten-inch wingspan, fluttering around on the floor.

"I picked it up and it started laying eggs."

She began an insect collection, an act that would lead to her peculiar work in New York today.

"I was a tomboy big time. My pets were waterbugs."

By high school she'd built an impressive collection, which she presented at the interschool science competition, all her dead samples neatly labeled inside little boxes wrapped in plastic, with explanations on the side.

"One of the judges at the competition was a museum curator. She let me start hanging out there. I learned how to collect insects and study them."

Her fascination grew when she accompanied a team of foreign scientists into the countryside to collect mayflies.

"Normally, in Sri Lankan culture, a girl doesn't go along with males alone, right out of high school. But then a girlfriend of mine, who was 23, ended up being chaperone."

She spent a happy month ranging over the forests of Sri Lanka. During the days she pinned mayflies. At night she sat by rivers, holding white sheets under black light. Thousands of mayflies landed on the sheets and Varuni caught them.

"This made me realize what I wanted to do for the rest of my life."

Soon the facination narrowed to one particular flying insect. She was surveying wildlife in the rainforest for a Sri Lankan environmental group when she gazed into a fluted pitcher plant and saw what for her was an incredible sight.

"Mosquitoes and mosquito larvae. I started reading about mosquitoes. There were 3,000 species in the world."

She dove into mosquito work at the Southeast Asia Cooperative Environmental Program after college, then worked on a World Health Organization study of vector-borne diseases. She discovered two new species of mosquitoes, wrote to the Smithsonian Institute in Washington for funding to study them, and ended up winning a fellowship that brought her to the U.S. Since completing a PhD at the University of Maryland, she'd been working at New York's Museum of Natural History, flying around the world, catching mosquitoes and studying diseases they carried.

Which is why museum officials had routed to Dr. Kulasekera the New York City Health Department's anxious call about a possible mosquito problem in Manhattan. Residents of streets in the East Nineties were complaining that even in December, mosquitoes were massing and biting them at night. When they killed the mosquitoes, callers said, the smashed insects smeared their walls with blood.

One city councilman had asked residents to send Varuni the insects, but the ones that came turned out to be crane flies, and "Crane flies don't bite. They don't show blood when you kill them. I was going to have to see for myself."

So now, in the middle of a snowstorm, she was heading up FDR Drive, carrying her usual mosquito-collecting net and sample cases on the front seat. Equipment she usually used in the tropics.

"I was thinking about global warming. Because northern areas are becoming warmer, many mosquitoes usually found in southern areas, like culex mosquitoes, are moving north."

And culex mosquitoes, Varuni knew, carry diseases that northern mosquitoes do not.

Like West Nile Virus.

By 1998, she was not the only one worried about tropical diseases moving north. The Harvard Medical School published "The Health and Economic Consequences of the 1997/1998 El Niño and La Niña," which warned, "We have begun to see the profound consequences climate change can have for public health and the international economy."

The report documented "anomalous climatic conditions during the 1997/1998 El Niño" and found them matching up with outbreaks of malaria in Colombia, Rwanda, Sri Lanka and the Punjab region of northeastern Pakistan where "the risk of malaria epidemics increases fivefold during the year following a major El Niño"; with cholera following flooding or droughts in Kenya, Somalia and Honduras; with hantavirus in New Mexico; with pulmonary syndrome during the Indonesian fires; with dengue in Cuba, Brazil and northern Venezuela; and with outbreaks of many pests, like medflies, which threatened Florida's $53 billion agricultural industry.

Florida's agricultural commissioner was quoted in the report. "We're like a state under siege."

The report's editor, Paul R. Epstein, became visibly agitated when the subject of tropical disease came up. He considered the danger urgent. Once a skeptical reporter asked him how researchers could ever be sure, even if diseases like malaria *did* show up in U.S. cities, how global warming contributed to the spread. After all, big U.S. cities had international airports, and infected travelers could arrive from the tropics, already sick.

The reporter used New York City as an example. "How can you be sure, if a New Yorker gets malaria, that an infected person didn't just fly in, and then a local mosquito bit him, and carried the virus to the next person it bit?"

Epstein explained that although the question sounded logical, the scenario was impossible because of the way mosquito-borne diseases are transmitted.

"A mosquito that bites an infected person is not instantly contagious. Before malaria becomes contagious the parasite has to incubate in the mosquito, and migrate from the mosquito's intestinal tract to its salivary glands. The parasite, while traveling, evolves from noninfectious to infectious. At seventy-degree temperatures it takes about twenty-eight days to migrate, but mosquitoes only live for two weeks, *so the disease never gets transmitted.*"

But as temperatures warm, Epstein warned, the parasite's metabolism speeds up. At 80 degrees, the mosquito might become infectious after only 10–12 days, not 28.

At 90 degrees, it becomes dangerous even faster.

"The other reason heat makes disease go up is that in warm, wet conditions mosquito breeding goes up," he said.

And as Varuni Kulasekera knew, in New York, in 1998, there had been plenty of warm, wet conditions. Although it was snowing on

December 23 as she left FDR Drive and drove to East 91st Street, temperatures the day before had reached 50 degrees, in a year that would be the hottest on record.

Police had blocked off the street with wooden sawhorses. They moved the blockades aside so Varuni could steer her Altima through.

She was dressed in black jeans, black rubber-soled boots, a black hat, black jacket, gloves and a multicolored silk scarf. She had a silver nose stud and thick black hair cut very short, matching her black eyes.

"People tell me, you don't look like a scientist."

As the neighbors watched, health officials clustered around a manhole, and workers used iron hooks to lift off its heavy round cover.

They stepped back in amazement when clouds of mosquitoes erupted out through the falling snow.

"They were culex mosquitoes."

For the next three hours, the workers opened more manholes in the area, and Varuni collected mosquitoes.

Every time a manhole was opened, the same thing happened. "There were thousands upon thousands of culex mosquitoes down there. Bells were ringing in my head. I told the officials that next summer was going to be bad."

Meaning, she warned, that in the summer of 1999, New York City might actually experience an outbreak of mosquito-borne tropical diseases.

She would turn out to be right.

West Nile Virus would kill 7 in New York in 1999 and infect 62. July would find Varuni in Queens, epicenter of the outbreak, traipsing through backyards of private homes the way she used to move through rainforests, looking for mosquitoes. She'd find thousands of larvae swimming in flowerpots, inflatable swimming pools and discarded tires in vacant lots. She'd take the samples back to the museum, needing a protected lab to examine potentially lethal insects gathered not from a jungle but close to New York bus and rail lines and highways, all within 40 minutes of Shea Stadium, Broadway theaters, Fifth Avenue shops.

"I stressed to the health department that the city will need a mosquito surveillance program over the next five years. To correlate weather, climate and environmental changes. To monitor mosquito-borne viruses."

The worst moment that summer would come in a beautiful back-yard, she'd remember, in a wealthy neighborhood within view of Long Island Sound. With a researcher from the Centers for Disease Control in Atlanta and a health department escort, Varuni would be asking the owner questions, jotting notes, when the woman's son would come out of the house and tell his mother: "That was the hospital that called. Dad's dying. Make the funeral arrangements."

Only then would the woman break down and tell Varuni something she'd been holding back as too upsetting.

Her husband had caught the West Nile Virus.

Varuni would never forget it.

"Her husband died that day."

North Carolina ... Delhi ... Washington

THE DANGER WAS REAL, Tom Karl finally believed. By March 1999, as scientists from around the world converged on Asheville to work on the now scheduled for 2001 IPCC report, the former skeptic had decided that their previous warnings had not been strong *enough*.

"It looks like the man-made greenhouse effect is a real significant global issue that's not going to go away," he said, host of the meeting this time, not just a participant, and chair of the Task Group on Climate Change Indices for the upcoming report.

Nineteen ninety-nine would turn out to be the fifth warmest year since 1880, Karl would report, despite the cooling effect from La Niña (the opposite of El Niño) in the Pacific. In fact, *every* year of the 1990s had been among the 15 warmest on record.

"No matter what we do, we're not going to be able to escape this unless we come up with some great new technology to reduce greenhouse gasses," Karl said.

He'd been converted by the massively growing body of evidence pointing in the same direction.

"There were new data sets on extremes and on oceans from China. Data from Europe, and from Russia. For the first time the Africans were playing a role."

New findings about the role of oceans in climate, programmed into models, had made them more accurate.

New upper-atmosphere data, from projects like ice station SHEBA, had improved models too.

"In order to understand the *whole* climate system we have to look at the surface *and* the upper atmosphere," Karl said. "Otherwise models could give an answer that looks right, but for the wrong reasons. For instance, we know earth's surface is warm but the stratosphere is cooling, so if a model doesn't replicate both, its physics is incorrect, which would mean its projections about the future are based on false physics."

But the studies all seemed to be validating each other. By 1999, Tom Karl knew, Michael Mann had extended his ice core and tree ring work to show the decade was the *hottest of the millennium*. Also, extreme precipitation events were "rising at two to three times the mean during a period when overall climate was changing faster than at any other time."

Karl had risen to become director of the National Climatic Data Center, the world's largest archive of weather information. He sat at the apex of the multi-million-dollar federal weather data gathering system, heading a staff that stored and processed surface, marine and upper-air information—everything from cloud and heat statistics to hours of sunshine per month—gathered by the National Weather Service, U.S. military, FAA, Coast Guard, volunteers, satellites, radar, remote sensing systems, aircraft, ships, radiosonde, wind profilers and rocketsondes.

"We like to think of ourselves as the nation's scorekeeper when it comes to climate," he said.

He'd even become a regular visitor to the White House, where he and other climate experts gave briefings to Al Gore. One time he joined President and Mrs. Clinton and Vice President Gore at a seminar at George Washington University, where he gave a ten-minute talk on extreme weather, broadcast over the Internet.

"The last five years have broken all previous records by a large margin," Karl said.

The constant work had taken a toll on his marriage. Within a year he'd separate from his wife, who would tell him, "this workaholic stuff had something to do with it. You're gone so much of the time, and when you're here, you're focusing on other things."

More criticism came from skeptics who had once praised his objectivity and now accused him of "crossing over," of being enticed by prominence, of being Al Gore's "buddy," even though the men had never been alone in a room together.

Karl shrugged when asked about it. "Look, we're trying to be cast as the honest broker, whether Gore is there, a more conservative president, or a more liberal president. If it's cold, we're going to tell you it's cold . . . We're going to tell you the scientific information we have."

And the information worried Karl. When asked over lunch in 2000 if global warming might have affected, *in any way,* the tornado that struck Nashville, Tennessee, in April 1998—and killed Kevin Longinotti—he began his answer with familiar caution.

"It's too early to say for sure. We haven't studied tornadoes enough."

But then he added, surprising the reporter, "I would be surprised if there were no effect . . . Tornadoes, like every other aspect of climate, could be affected."

The reporter asked if warming might have played a role in hurricanes Andrew and Mitch, and in European storms like the one that hit Emily McDonald's school in 1990.

Yes, Karl said.

"Hurricanes will probably be more intense, wind speeds five to ten percent more intense. A gradual increase. A number of models are suggesting now, and it makes sense from what we know about the effect of a warmer world, that hurricanes would generate more precipitation."

The reporter asked if global warming could have encouraged the Yellowstone fires of 1988, the Oakland fire of 1991 and the big New Mexico fire of 2000.

"As conditions dry out," Karl affirmed, "and as it gets warmer they dry more quickly, that's one of the things you'll get to see . . . Again, you can't pin any single event on warming because natural variability has an effect on all these things, but what's happening is that on *top* of natural variability you have these trends. So when it's warmer, it's a *little bit* warmer, so it dries out the soil faster. A *little bit* faster."

How about the deadly Chicago heat wave of 1995?

"Heat waves become more frequent . . . In the Chicago heat wave of '95, you had very elevated temperatures through the night. When people think of heat waves, they think in terms of hotter than heck in the daytime, but the *night* is much warmer than it used to be. You don't get much relief."

Could the Sudanese drought have been connected to warming?

"Yes. Increased warmth tends to make areas that don't have rain dry out much quicker."

The bottom line, Karl said, was that, "there are not *any* events that are not influenced at all by the greenhouse effect . . . With a long-term perspective, any particular weather event we look at now would have less severe consequences if the world weren't warmer than it was a hundred years ago."

Many of the IPCC scientists agreed with him, and leaving Asheville, they got to work. By autumn 2000, their draft report would say that human activities had "contributed substantially to the observed warming over the last 50 years," and predicted that by the end of the twenty-first century, average global temperatures might rise much *higher* than originally anticipated, as high as 11 degrees Fahrenheit above 1990 levels. By comparison, about 20 degrees separated average temperatures between 2001 and the last ice age.

In short, the IPCC was now saying that due to human-caused greenhouse effect, temperature swings that would otherwise require thousands of years to occur may quite likely now happen in a century.

In short, with better models, better records and more complete studies, work which critics of global warming gasses had demanded in 1988, the worst-case scenario *had just gotten worse*.

For Karl, the big remaining uncertainties did not even relate to global questions anymore but were "at the regional level. What's going to happen in different regions . . . Will we see more storms in *my* state? *my* region?"

At the lunch, Karl, a careful eater, ordered a hummus plate. He wondered whether, as various models suggest, the climate of Illinois would warm to resemble Oklahoma's during the next century, or eastern North Carolina's.

"The difference is how much precipitation falls in the summer."

He worried about ecosystems. "The ones at risk are the ones we know the least about . . . You could have ecosystems totally disappear. You could envision certain pests becoming prolific as their enemy diminishes . . . That could be the story for the next century . . . All the conditions are right for some major events."

In short, "You could end up with big surprises," said the man who had been watching weather surprise him ever since he was a boy, and who a decade earlier had tended to believe that future climate surprises would probably mitigate greenhouse warnings, not enhance them.

"That's quite a wrap-up from a guy who used to be a skeptic," the reporter said.

Karl mopped up hummus with pita bread. He ate some salad. He drank some juice.

"We all change."

———

By 1999 the search for solutions was quickening, and early one morning that autumn, hoping he'd found one, Dr. Ajay Mathur pushed his way through Delhi's main train station, past a clamorous mass of business-people and beggars. Lepers held out stumpy palms, begging for alms. Brightly turbaned, bearded Sikhs thronged pedestrian bridges, heading to boarding platforms. Young pilot-cadets for the Indian air force awaited trains beside aging hippies bound for Nepal, orange-clad Jain priests making pilgrimages to Benares, tourists carrying bottled water from Delhi hotels.

Mathur took a seat in a first-class compartment of the Agra-bound express, barely able to contain his apprehension over whether his four-year gamble would work out today. The destination for the sensitive, hardworking engineer, three hours southeast, was a small factory inside the "Taj Mahal Trapezoid," as the Indian supreme court had designated the area around the tomb.

Pollution there was so acidic that the Taj was deteriorating. To save it, the court had ordered local factories to switch from coal to gas power, or close—a ruling that had set the stage for Ajay's test today.

Thirty-seven-year-old Mathur had downy black thinning hair above tortoiseshell glasses. He radiated passion for his project and hoped it might one day keep tens of thousands of tons of carbon dioxide from reaching the atmosphere annually—from the third world.

The train lumbered through a vision of poverty and energy deprivation—the shanty slums of Delhi—and into the chilly countryside, over rickety railroad bridges and past villagers collecting cow patties or hauling wood for cooking, heating and cottage-industry power.

And power had occupied Ajay Mathur's daydreams even as a child, when he'd accompany his father, an irrigation engineer, to hydroelectric dams, and watch him stop the family car on outings, walk across fields and ask farmers, "Do you have enough water? *Do you need more energy?*"

Ajay had chosen a career in engineering too. To him, engineers were the inheritors of a vast patriotic responsibility. "In Indian history, one of the most important people in every court, as far back as 200 B.C., was the Superintendent of Canals. Every emperor had one. If canals deteriorated, it caused a huge exodus of people."

A steady supply of energy to India would prevent similar disruptions in modern times, Mathur believed. But he feared global warming, and wanted that energy to be clean.

To further his dreams he had graduated from India's most prestigious engineering school and had studied in the U.S. Now he worked at TERI, the energy and environment institute, where he focused on the tens of thousands of small rural factories that fueled a huge segment of India's economy as well as chunks of China and the third world.

"The rural energy sector is growing faster than anywhere else and will cause our greatest rise in greenhouse emissions over the next twenty years—as high as thirty percent," he worried. "It's what we're trying to prevent."

The supreme court ruling had given Mathur his chance. Hundreds of small factories were clustered in towns near the endangered Taj, their owners desperate to stay in operation, resentful at being ordered to cut emissions, furious at con men who had flooded the area, selling bogus contraptions to "cut down on pollution," and suspicious of urban academics like Mathur who wanted to help.

"They didn't trust us at first."

Then one factory owner in the area had met Mathur and worked out a deal. The owner, Mr. Islam, would supply labor for Mathur's staff to design a low-emission furnace. If it worked, Islam would pay for it. If not, it would be free.

"All the factory owners were watching to see what would happen,"

Mathur said. Now his train arrived at Agra, where he was greeted by Mr. Islam, a lean, healthy 60-year-old. In Islam's van they proceeded southeast along a trunk highway, past small farms and villages within view of the Taj Mahal. "It's so beautiful," Mathur the patriot breathed.

They entered Mr. Islam's town of Firozabad, once forest, now a road lined by 150 factories: brick, two-story buildings with steel pipe chimneys belching dark smoke.

The van passed through a black iron gate onto a dirt courtyard, stopping before Islam's home—two stories high with large windows, and mango trees planted alongside.

Walking toward the factory, Mathur recalled what he'd seen inside on his first trip: an eight-foot-tall brick beehive-shaped furnace centering the high room, with flames leaping inside, as two dozen workers in shorts and tee shirts, sweating rivers, twirled long tubes of metal with liquid glass bulging from the ends. They fashioned bangles or lightbulbs from glass glowing red, green, yellow, blue.

Mathur, transfixed, had thought, *There's so much heat being wasted here*—and all over the third world.

That day Mr. Islam had waited while Mathur took notes and asked questions, and then, over delicious samosas, cakes and kabobs, had politely asked, "Can you help?"

It wouldn't be a huge modern project, not like the mammoth dams on which Mathur's father had worked, not like the kinds of solutions to global warming needed in the U.S. and the first world. But it applied to hundreds of thousands of small businesses, and the TERI staff threw themselves into getting funds from a Swiss NGO to design a new furnace and confronting a wide range of obstacles to switching from coal to gas in a rural cottage industry.

The job was actually dangerous. Gas rose as it warmed, so if you replaced the fuel source at the bottom of the furnace, the roof would collapse from pressure. That had happened to the redesigned furnace of Mr. Islam's neighbor. So Mathur's staff suggested that *his* furnace pump gas in at the top, not the bottom. That way, instead of rising against the roof, it would be forced downward by the flow.

"Mr. Islam was skeptical," Mathur recalled.

The staff then designed an air circulation system to recycle the hot air in the furnace after initial use.

"We went through many ideas that did not work first.

"We increased the size of the dome, so the arch could better with-stand heat stresses. The Swiss paid for a British glass manufacturing expert to come in and give advice."

Finally, in late 1999 the project was finished. When Mathur walked into the factory now, he saw two furnaces: the old one, still working, and the new one, not yet used.

"Fire it up," Mr. Islam ordered his men.

It worked.

"I was ecstatic," said Mathur, whose project had come about not because of global warming, but because local needs dovetailed with international ones. The seed money had come from a country worried about warming. The technology would pay for itself. The patents would be given gratis to manufacturers in the interest of environmental progress.

As far as Mathur was concerned, projects like this one, and cleaner kilns from China, biomass gassifiers "to efficiently burn wood," solar ponds and wind power constitute practical ways that poor countries can meaningfully participate in greenhouse gas cutbacks.

"Foundries—brickmaking, pottery, dying cloth—produce forty per-cent of our industrial pollution, and in other developing countries, the percentages are even greater."

By 2001, Mathur would be getting lots of calls from factory owners across India who wanted to try his furnace.

"That little furnace multiplied by several tens of thousands, globally, will remove how much carbon dioxide? A meaningful number," Mathur said proudly, back in Delhi.

"It will be a great day for me when that furnace lights up in Cairo, Africa, Bangladesh, Pakistan," Ajay Mathur said.

———

With the Kyoto Protocol languishing in political limbo in 1999—unrat-ified by any major country—its enforcement issue and tradable emis-sions issue still undefined; and with a treaty opponent, George W. Bush, running for president in the U.S., it was becoming clear that the treaty might never go into effect, and the search for alternative solutions in the developing world became a patchwork of hopeful national, corporate and individual initiatives.

In London, Deputy Prime Minister John Prescott sat down with representatives of major British industries and designed a national climate-change tax to go into effect in 2000.

"We made it clear to them we would impose the tax if necessary, but if we could come to a voluntary agreement we would."

Under the plan, tax revenue would recycle back to business through lower welfare contributions and rewards for energy efficiency. British utilities would be required to produce 10 percent of their energy from renewable resources and gasoline taxes would jump by a huge amount.

In mainland Europe, the German Parliament passed the Ecological Tax Reform Act, increasing the national gasoline tax in steps. Other new legislation required more energy-efficient building codes and taxed electricity, while offsetting costs with tax reform for the poor.

The Danish government introduced a "green tax" and ordered a reformation of the national electricity sector, tradable emissions reductions and a CO_2 emissions cap.

Nine other countries, including Switzerland and France, instituted energy taxes, while others offered tax credits to coax companies or homeowners into greater energy efficiency. Australia credited rail and household improvements. Canada credited equipment to capture flared gas. France changed amortization rates for energy-efficient materials.

In the business world, two old antagonists, Jeremy Leggett of Greenpeace and Rodney Chase of BP Amoco, ironically found themselves competitors in selling solar energy systems. Leggett, disgusted with the diplomatic process, had started the "Solar Century" company, which made solar roof panels and lighting systems. Prices were prohibitively high due to production costs. But he'd installed a system in his own home and found that despite England's cloudy weather, it provided all his electricity as well as extra energy sent back into the public grid.

Chase, at BP Amoco, was also steering investment to solar energy—not shying away from oil exploration, but company officials also planned for a world where fossil fuels would play a decreasing role.

"When you look at energy ten to fifteen years from now," said Chase, "we are moving away from oil, toward natural gas, and toward even cleaner forms of energy. We've gone from wood to coal to oil, natural gas, and what's beyond that? For us, solar-generated power, photovoltaic energy, converting photons to electrons . . . We are driving the market's creation, not following it."

What he meant was, BP Amoco planned to install solar energy systems in their 30,000 gas stations around the world, to demonstrate their effectiveness. But the only way to do that was to build new factories to supply themselves—plants that would go idle if no new customers were found.

By late 2001, Chase said, U.S. drivers entering BP stations would find themselves gazing up at solar canopies providing all power in the station with the exception of some energy in the maintenance shop. The system would be available for purchase for private homes.

"If we can make similar R&D breakthroughs," Chase said, "far more of the surface of the buildings we work and live in—other building materials—will be generating electricity one day. I know this sounds fanciful and dreaming. I recognize we're talking a long way ahead."

As BP worked on the fuel side of the equation, on the auto side, Honda unveiled its new two-seat "Insight," designed with a "hybrid" engine that ran on gasoline and electricity.

The Insight won a Sierra Club prize for "excellence in environmental engineering" and got 70 miles per gallon on the highway, 61 in the city. It looked like any Honda inside, although its futuristic aerodynamics drew stares. It started up silently. Electricity ran the car up to 20 miles an hour. Then gasoline smoothly kicked in, and the engine and braking system automatically recharged the battery so owners never had to do it, as with fully electric cars.

Sticker price was $20,800 with air conditioning. Toyota planned to release a five-seat hybrid in 2001. Ford, producer of an all-electric car, studied ways it could be systematically used to move commuters between home and public transport. Other companies, including DaimlerChrysler, GM and Toyota, invested hundeds of millions of dollars in fuel cell technology, which used cleaner hydrogen to power cars.

Writing in *Time* magazine, author and environmentalist Mark Hertsgaard called for the U.S. government, purchaser of 56,000 vehicles a year, to buy hybrids or fuel cell cars. He said government pump priming had been responsible for America's computer boom because many technologies fueling it had been developed with Pentagon subsidies in the 1960s.

Hertsgaard also argued that by helping developing countries buy efficient technologies, industrialized nations would reduce global warming and create jobs back home.

With a growing sense that Kyoto was failing, more suggestions for solutions popped up in magazines, on TV shows, in congressional offices.

Bill O'Keefe of the American Petroleum Institute urged the White House to revise depreciation rules to encourage the retirement of energy-intensive equipment, and to expand scrappage programs to reduce emissions from older vehicles.

Rafe Pomerance—the man who had first brought the climate issue to Tim Wirth and who had even led the U.S. negotiating team in Berlin at one point, had left the State Department and now lobbied in Washington for "Sky Trust," a plan that he claimed would reduce greenhouse emissions 7–10 percent right away without harming the U.S. economy.

The issue was so emotional he grew agitated talking about it. For a long time he'd been tormented by the way the Clinton administration had handled the Kyoto treaty.

"I keep thinking about the mistakes."

Agreeing to legally binding targets without adequate preparation had been a mistake, he said. "Nobody knows what the costs will be. The models show a wide range."

But his depression had changed to enthusiasm with "Sky Trust," under which, Pomerance explained, costs would be capped, and corporations or utilities introducing fossil fuels into the U.S. economy would have to buy permits for the carbon in the fuel. The government would auction off permits for an amount equal to 1990 emissions. Permits would be tradable, and extra ones could be purchased for $25 a metric ton, a kind of unofficial tax.

"Don't use that word, 'tax,' " he snapped. "It's not accurate."

To offset the cost of permits, utilities and energy corporations would probably raise prices, Rafe knew, but the plan would compensate consumers directly by returning 75 percent of the receipts from permit sales to U.S. citizens in annual checks. They could use the money to pay energy bills, or they could cut energy use and keep the money.

❂ Under the plan, 25 percent of permit receipts would go to people like truckers, who had to drive long distances, or to especially hard-hit energy producers. And 25 percent of the *auction* revenue would ease the transition of firms, employees or households having difficulty changing to a low-carbon economy (like coal miners and coal companies).

Pomerance pushed the plan under the auspices of "Americans for Equitable Climate Solutions," funded by foundations.

By autumn 1999, he'd been working on global warming for over 20 years. On a cool morning, after a bagel-and-coffee breakfast, he stood on the porch of his home in Adams-Morgan, in Washington, asking a reporter for a ride downtown, where he had a Sky Trust meeting.

He was a tall, bald man, liked by allies and opponents, his passion endearing rather than off-putting. His long coat was open and, face red and long arms waving, he sounded like a character in a science fiction movie, trying to convince his neighbors that dangerous aliens had arrived.

The alien in this case was human-made carbon dioxide.

"Don't you *understand?*" he cried. *"It's gonna double!"*

Plan after plan. A final one came from Paul Epstein, the Massachusetts expert on tropical diseases, who, worried about malaria spreading in a warmer world, had joined with energy experts and policy specialists at the Harvard Center for Health and the Global Environment, to call for:

1. The switch of national subsidies from fossil fuels to renewable energy technologies in industrialized countries, and the retraining of coal miners.
2. The elimination of regulatory barriers against new energy sources (like laws permitting old coal-powered plants to operate) in tandem with progressively more stringent fossil fuel efficiency standards—so climate-friendly energy sources could compete in the marketplace.
3. The incorporation into the Kyoto Protocol of fossil fuel efficiency and renewable energy content standards. And the creation of an international energy modernization fund to buy energy-efficient technologies for developing nations.

Epstein, author Ross Gelbspan and the group called for financing the new fund through a tax on international money transactions, which total roughly $1.5 trillion a day.

"A quarter of a penny tax per U.S. dollar on 'close of the day' positions in the international currency markets would yield $200 to $300 billion," Gelbspan wrote. The tax had first been proposed by Nobel Laureate James Tobin, and was already under consideration in various countries, although not as a way of mitigating global warming.

Back in 1988, when Jim Hansen had warned that the greenhouse effect was here, millions of people had never heard of it. By 1999 universities offered courses on it. High school math teachers used fluid dynamics equations of global warming as teaching aids. Wall Street brokers attended lectures on El Niño, global warming and weather futures as investment.

As the millennium approached, local governments funded studies to assess regional effects. The California Energy Commission awarded a $2.2 million contract for an evaluation of potential impacts on the state. The Army Corps of Engineers studied ways to assist Long Island communities if more violent storm surges hit the area. Experts considered ways to keep surges from inundating low-lying runways at JFK and La Guardia Airports.

At the same time, Chicago, Los Angeles and Newark, New Jersey, joined an international federation of cities coordinating efforts to cut greenhouse gasses. New York City published guidelines for energy-light technology for public buildings, and the guidelines were followed in renovation projects including the Bronx Criminal Court.

Doctors, businessmen and even tourist writers speculated on warming. Would high temperatures kill off coral reefs from Florida to the Red Sea? Would the maple trees of New England migrate north as Vermont grew warm? Would tourist meccas of southern Europe become malaria risks?

The sad truth was, the speculations were necessary because greenhouse gasses were still rising. In October, the degenerating situation was summed up by Michael Zammit Cutajar, executive secretary of that year's

UN Climate Change Convention. By 2010, he said, instead of dropping 5 percent from 1990 levels, as the Kyoto Protocol had aimed, emissions would rise by 18 percent. The U.S., Russia, Japan and Germany were the biggest emitters, with China and India in the top six.

If you allocated tonnage by *person* instead of country, he added, the U.S. would emit 20 tons of CO_2 per person; China 2.5 tons; India, less than one.

In Washington, the White House still issued strong statements about climate change, but in a bitterly partisan Congress, even small initiatives to cut emissions were failing, with battles erupting over riders or directives stuck onto spending bills, as conservatives vowed to stop any move toward Kyoto targets.

For instance, one fight revolved around the increasing number of vans and recreational vehicles on U.S. highways. Classified as "light trucks," they operated under lower fuel efficiency standards. Each year, congressional Republicans attached a provision to the Department of Transportation's budget preventing the raising of fuel standards for vans and RVs—a provision that cost the country an extra million barrels of oil a day, environmentalists claimed, adding that U.S. vehicular emissions were increasing 2 percent a year, according to a study made at the University of California at Davis.

Real progress, in short, remained an illusion. As earth burned, Congress cut 8 percent of budget allocations for solar, wind and other alternative forms of energy. It blocked a White House initiative to slice $750 million from the federal government's annual energy bills. It shot down a proposal to grant tax credits for buying efficient air conditioners and rooftop solar units.

Global warming, Republican congressman Dana Rohrabacher said, was "liberal claptrap."

But to single out the U.S. was to ignore problems in other countries. Even the European Union, bullish proponent of the Kyoto Protocol, was having trouble meeting targets. Germany had promised to reduce emissions 21 percent below 1990 levels, and actually sliced them 17 percent, but the number primarily represented the closing or cleaning of dirty plants in former East Germany. In other words, progress had been easy, and now it would get hard.

"Despite strong political commitment, it will be difficult for Germany to reach its Kyoto targets," said a Pew report.

The Netherlands had aimed for 6 percent cuts below 1990 levels. By 2000, emissions would increase 17 percent.

Energy poorer Spain, as part of a "bubble" of European nations, had been allowed to *raise* emissions 15 percent. But the Pew report projected Spain would top that amount.

Only the United Kingdom had exceeded its target, reducing emissions by 14.6 percent below 1990 levels.

Meanwhile, with the world's population surging, its basic needs exploding, the vast hungry energy markets of the developing world had oil companies licking their chops.

As Exxon chairman Lee R. Raymond told the World Petroleum Congress in Beijing, "In fifteen years, this region's economy should almost double, shifting the global economic center of gravity toward the East . . . As we all know, economic growth and higher standards of living require energy . . . The countries with the highest . . . standards of living are also those with the highest energy use per capita."

Exxon wanted to supply that energy.

Raymond said, "We estimate that oil for transportation and industrial use in the region will grow by nearly ten million barrels a day by 2010. That's equivalent to about forty new large refineries over that period—three per year . . . I know there are some people who argue that we should drastically curtail our use of fossil fuels for environmental reasons . . . But let me state at this point my belief that such proposals are neither prudent nor practical."

Raymond received big applause.

And if gasses keep rising? Will day-to-day life change across the U.S.? "Tell me what you found out," a *New York Times* reporter asked the author of this book, with a smirk, after the research was completed. "So, ha ha, are we all going to die?"

No, the author said. We're not all going to die, just like we won't all die from cancer. Just *some* people will die, and probably, living in a wealthy country, one of them won't be you. *Probably* you won't be the one to catch malaria or dengue fever in Minneapolis, San Francisco, Baton Rouge, Bangor. *Probably* you won't be one of the people caught in a storm surge while stuck on a highway, fleeing a hurricane in Myrtle Beach, South Carolina, or New Orleans. You personally will probably not be sent, as a soldier or journalist, to an area of Africa, Asia or Central America destabilized after a hurricane or drought. Probably, as a resident of a wealthy country who doesn't live along a low-lying coast, or in a river valley, the worst you'll suffer from climate change will be a shift in where you take vacations, a rise in your tax bills to pay for rebuilding after disasters, a surge in insurance rates, possible water and energy shortages, and a few less animals or forests to watch on *National Geographic* TV specials, since they won't exist anymore.

The author told the journalist that he had come to consider questions like "Are we all going to die?" the way he thought of the old hit-movie *Jaws*, where a shark swam offshore a New England community, periodically eating people. In the end, the shark only ate four or five people out of thousands in the community, "so what was the big deal anyway, statistically?" the author said. "It was only a few people, and probably not you.

"Then again," the author added, "maybe, when you're sixty-five, and you get a heart attack on a hot night, it will never even occur to you that at three A.M., temperatures wouldn't have been as hot if the climate hadn't changed."

A hint of what day-to-day life in the U.S. would be like for many Americans in the future, the ones who would be affected but *not* die from climate change, if IPCC projections are right, was provided on Saturday, October 16, 1999, a calm, lovely morning in eastern North Carolina, when two men who had never met each other stood frowning at the balmy sky above their homes and considered how to protect their vastly different communities from a coming storm.

Bob O'Quinn, alderman of the town of Wrightsville Beach, was a prominent lawyer in the kind of barrier-island community in which millions of Americans aspire to live, a place representing the finer life.

Walter Scott, volunteer fire chief in the rural hamlet of North East, an hour inland, was top official in an unincorporated area that lacked resources to recover from disaster—especially after Hurricane Floyd had hit it a month before. Scott's men had saved 500 people that night from death.

Now a new hurricane, Irene, was taking her time meandering up the Atlantic, like a bully looking for a victim, staying in striking range. It was as if previous hurricanes had worn a groove in the atmosphere leading toward the Carolinas, like a deep rut in a road.

Between O'Quinn and Scott lay not only physical mileage but gaps in wealth and outlook. To watch them today would give a fair notion of how a worsening greenhouse effect could change life for millions across the U.S.

By October 16, North Carolina had been hit by five hurricanes in the last four hurricane seasons: Bertha and Fran in 1996, Bonnie in 1998, and Dennis and Floyd in 1999, when two evacuations had already emptied Wrightsville Beach. Fran had damaged Wrightsville Beach so badly the town had been off-limits to residents for a week. After that, with water and electricity off, they'd lived under curfew. National guardsmen had patrolled streets at night on bicycles normally rented by tourists, while helicopters directed blue searchlights at millions of dollars in wreckage. Roadblocks and marine patrols had kept away potential looters. And for months, even after essential services resumed, residents had woken each day to the cacophonous noise made by roofers, debris haulers, insurance adjusters at front doors and bulldozers rebuilding beaches.

"It's as if God set up a giant target out in the Atlantic Ocean and told his storms to try and hit it," one Red Cross official said.

The hits had been dead center. The eyes of Fran, Dennis and Floyd had passed directly over Bob O'Quinn's house.

Today, O'Quinn and his fellow aldermen had to decide once again whether to evacuate.

"It's especially tough on the elderly."

He was sick of the storms, the waiting, the evacuations.

There'd been so *many* evacuations that his neighbors knew procedures by heart. Leave in daylight. Take small valuables and important papers. Travel light. Shut off electricity before going, and board up windows. Carry lawn furniture inside. (At beachfront hotels, clerks threw them in swimming pools to keep them from blowing through a house, window, car.)

A few of O'Quinn's acquaintances had moved away after so many hurricanes, but for most of the 3,000 citizens of wealthy Wrightsville Beach, and 32,000 summer visitors, the risk of staying was still—in 1999—worth the reward. Life was pleasant at the beach and in Wilmington, the city across the inland waterway. Although they had separate governments, the cities shared an easygoing lifestyle including golf, tennis, theater, symphony in Wilmington, marinas, a large campus of the University of North Carolina and a movie studio, biggest on the East Coast, where films including *The Hudsucker Proxy* had been filmed, and the television show *Dawson's Creek* was shot.

Some of the actors or directors working at the studio rented or owned houses on Wrightsville Beach, as did retirees, university students, old-line Carolina families and an increasing number of full-time young families moving in. On average days, the white sand beaches were dotted with joggers and dog walkers. Bottlenose dolphins abounded offshore. The channels, estuaries and Cape Fear River were filled with boaters, fishermen, kayakers and sun lovers, including boating and sports fishing enthusiast Bob O'Quinn.

Fifty-three now, he'd loved Wrightsville Beach since boyhood. His grandfather had owned a vacation home here. His doctor father had moved the family to the town full time. After college, a stint in the air force and law school, O'Quinn had happily returned.

He was a large, handsome man, an expert in business and employment law who wore suspenders at home and the office. He enjoyed coin collecting and, now and then, a trip to a Bahamas casino. His compassionate streak manifested itself in a tendency to extend himself for needier clients, occasionally taking on their legal problems for free. His first thriller, *The Bermuda Virus*, had just been published.

Between his roots in the community and affection for his town, he keenly felt the trauma of each evacuation. Homeowners never knew when they left if their home would be there when they returned, or in what condition they would find it. And even in a community wealthy enough to adapt to trouble, public workers might spend days blocking bridges, keeping people off the island, and months being screamed at by citizens frustrated by the pace of recovery.

"Each time we evacuate, public works secures the infrastructure. They pump the water tank full of water, empty sewage lines and shut off valves. We have a physical notification process where someone knocks on every door. Evacuation is voluntary at first, then, with eight hours to go, it's mandatory. The town becomes a ghost town. Some people go to Wilmington. Some go to hotels, or inland. The Board of Aldermen goes to a predetermined location where we stay in session until the emergency is over."

Evacuate or not? he thought now.

"It's an easy decision if it's blowing hard, right down your ass. This one was harder."

O'Quinn sighed and mounted the stairs to his house, one of a new generation built to withstand hurricanes. It's the kind of home that more coastal dwellers will need in a post-greenhouse world. The two-story wood and gray-shingle structure was raised on pilings lifting it 14 feet above the floodplain. O'Quinn's two family cars were parked beneath.

"My wife, Catherine, lost her cranberry-red Jaguar to Hurricane Fran. She lost her Nissan to Hurricane Bertha."

Also on ground level was a small room with breakaway walls, shielding the washer and dryer.

"The Maytag man is under standing orders. After each hurricane he brings a new washer and dryer. That's happened three times, starting with Fran. Wait. Was Bertha first or Fran? Why don't I have my hurricane list?"

Reaching the landing, O'Quinn heard the Weather Channel inside, which was always on when a hurricane was coming. He entered a side door into the dining/living room, filled with a long table, comfortable furniture and bookshelves, flooded with light streaming through five panels of sliding glass, outside of which was a wooden wraparound deck.

"Storm shields were made to cover the windows. Tile floors were planned for water to blow underneath. There are no windows on the north side of the house except for a tiny bathroom window. After these hurricanes I walk under the house, turn on the hose and wash away marsh mud. I call the Maytag man. The AC's on. Everything's fine."

But everything was not usually fine, he knew, for his friends and for his parents and brother Ted, whose nearby homes were *not* built on pilings. Those homes had required months of cleanup after hurricanes Bertha and Fran.

Despite the strain, though, as he phoned the city manager, O'Quinn refused to believe that the storm glut was more than coincidence. Hurricanes had always struck Carolina in cycles—a bad one in the fifties, a quiet one until the nineties. He believed the hurricane spate would subside. He believed Wrightsville Beach would prosper. He believed that with new building codes, better forecasting and the inevitable swing of the natural cycle, storms would be overcome, and the city's attractions would continue to compensate for any hurricane risk.

Today only the near future was on his mind, though. His job was to stay in contact with police and fire officials, who in turn were checking with county emergency officials, who were calling state officials, who were phoning the National Weather Service.

While he was on the phone he heard snatches of conversation from the kitchen, where his wife, Catherine, chairwoman of the Wrightsville Beach Centennial dinner dance—a celebration of the town's 100th birthday, unfortunately scheduled for tonight—was making contingency plans. She had been arranging the black-tie affair for a year, designing invitations, selecting a band, poring over menus, choosing the site—the Wrightsville Beach Holiday Inn resort, back in operation after being damaged by Hurricane Fran.

"The dance was supposed to be a celebration of the town's birthday, but also of our recovery from the last big storm," said O'Quinn.

The dance was still on at the moment but even if it were held, with Irene approaching, would the musicians reach the town? Would the storm speed up and arrive early? Would celebrants be discouraged if they missed an event planned partially to get over the last?

"We just didn't expect to get hit by number six," lamented O'Quinn.

Number five, Floyd, had been bad, slamming into the coast on September 16, only days after Hurricane Dennis dumped torrential rains upon the state. Coming ashore over Bob O'Quinn's house, Floyd had dropped over 20 more inches of rain on Wrightsville Beach, slightly less inland—causing a 500-year flood that swelled rivers and inundated towns from Rocky Mount to the coast. It closed sections of 300 roads, including swaths of Interstate 95. It robbed 1.5 million homes and businesses of power. It killed 48.

Over 110,000 hogs had drowned in the floods, and over a million chickens and turkeys. Their carcasses joined sewage, industrial waste, garbage and raw chemicals washing down the Tar, Neuse and Cape Fear Rivers to the coast.

The truth was, even now, weeks later, as Bob O'Quinn arranged to meet the other aldermen this afternoon at Town Hall, he could see Floyd's handiwork by raising field glasses and gazing out at Banks Channel. The normally blue water had been brown for weeks, with wavelets leaving yellowish froth on the beaches. The waterways, normally filled with boaters, were relatively deserted. Surfers had been falling sick, contracting fevers, sores, swollen tongues. They reported finding dead hogs in the surf off nearby Carolina Beach.

Just that week, O'Quinn knew, researchers from the University of North Carolina at Wilmington had been crisscrossing the tea-colored Cape Fear River plume—the "fresh" water that washed out to sea—measuring it for toxins, counting sea birds, trying to estimate damage.

The good news, before Irene headed toward the Carolinas, was that slowly the area off Wrightsville Beach was clearing, although researchers feared that runoff might create a huge dead zone further north, in Pamlico Sound.

"We're seeing an ecological event on the catastrophic scale," fretted Dr. Hans Pearl, a marine scientist at the University of North Carolina at Chapel Hill.

Now 5–8 more inches of rain were expected. In the capital, Governor Jim Hunt was preparing to declare a state emergency even while he was still in the process of trying to get funds to clean up after the last storm.

O'Quinn and the other aldermen decided to make the evacuation decision when they had more information.

"I absolutely *hate* this," said Bob O'Quinn.

———

Walter Scott's community of North East faced fewer options that day, boiling down to wait, hope and pray. If in the wealthier community of Wrightsville Beach, in more relaxed moments, Bob O'Quinn could explain how, in the long run, an occasional hurricane provided an infusion of money into a community, North Easterners had a less esoteric view.

Hurricanes Dennis and Floyd had turned Walter Scott into a fire

298 · BOB REISS

chief without a firehouse, a homeowner without an intact home and a family man who'd only recently managed to return his wife and two daughters into the barely livable wreck that had been their house a few weeks before.

Located along state route 41 in rural Duplin County, North East looked so insubstantial to passing drivers that they never knew it had a name. No signs identify the scattered, rolling collection of single-story homes set back from the road, the feedlot, soybean fields, community hall, wooden supper house called Pinks, and small grocery and, far back along dirt driveways, scattered turkey houses.

Only the low redbrick firehouse, visible for a fraction of a second, says North East.

On the morning of October 16, though, passersby slowed in shock when they saw the community. The highway was a gauntlet of wreckage. Volunteers outnumbered residents. In the parking lot of Pinks supper house, flooded and closed, buses disgorged gaggles of aproned men and women, helpers-for-the-day from churches around the state. They donned long rubber boots, orange vests and face masks against fumes, and tromped in groups into soggy houses to rip up mildewed walls and carpeting, to shovel out sludge filled with turkey waste, to haul ruined couches and appliances to the road.

Tractors pulled trash away, but the piles only grew.

Trucks passed, loaded with drowned turkey carcasses.

A couple from the "Baptist Handyman Ministry," out of Jacksonville, North Carolina, was stopped at one home, "helping others in Jesus's name," as the logo read on their specially designed trailer filled with wheelbarrows, hoses, rakes, air guns, shovels, ropes for pulling trees off houses, sledgehammers for busting through mildewed walls, hoes for scraping up flood-destroyed floor covering, shovels for removing mud from houses and chain saws for cutting up knocked-down trees.

"According to the Bible, there will be more storms as we get closer to the time of the Lord," said Andy Wood, pastor, handyman, ex–tool and die worker now in service of the Lord. "This is a fast-growing ministry, started because storms are growing more frequent. We had Bertha and Fran here, Hugo and Andrew further south. Floods in Kentucky. Tornadoes are more frequent than they used to be. We used to go years without one."

Certainly, as more storms had hit the Carolinas, *why* they kept coming had been a subject of church sermons. Some said the Lord was coming, or the country had lost sight of God. Less religious people saw
counselors in increasing numbers, their troubles exacerbated by the
unrelenting storms. Why me, congregants asked ministers, patients asked
psychiatrists, neighbors asked each other. Why us?

At North East, meanwhile, at the community hall, a Salvation Army
lunch wagon dispensed sandwiches as Red Cross workers interviewed
victims to determine aid eligibility.

The visitors sought out Walter Scott if they needed a signature on a
request form, a means of finding a person, or directions to a house. Walter's name constituted official sign-off for the Red Cross and the state
government. There was no mayor in North East, no council, no police.
There was just stocky, curly-haired Walter, 38, with his goatee, rectangular wire-rimmed glasses, bowl-cut hair and sand-colored sideburns
going gray at the edges. He chain-smoked thin cigars. Normally he
worked as a superintendent of maintenance at a hosiery mill.

On the night of September 17 he'd been awoken by a 2:00 A.M.
phone call from the County Sheriff's office, which had received a warning of water rising in a nearby trailer park.

Walter drove out to check. "I made it to the feedlot on route 41.
The water was three and a half feet across the road."

Like all North Easterners, he knew the area flooded every 20 years or
so, but those floods were four inches high, not like this. Already, on the
watery highway, he saw slow-moving headlights from the dumptruck of a
neighbor, Gary Cantrell, who'd been woken by his flooded-out brother,
and had single-handedly started rescuing people, driving along like a Paul
Revere of Hurricanes, blowing his horn, gathering up stunned residents
who tottered from their houses. He took them to the firehouse.

Soon the road was so flooded Cantrell had to steer by memory. Road
signs were underwater. Bridge railings were underwater. Fire ants were
taking refuge on people's porches, swarming over them as they came out.

"They were eating one old lady alive."

With the volunteer firefighters living close to the station, the department had won numerous awards for response time. Within 20 minutes
the men were on the scene, and Walter's wife was on the phone, waking
neighbors, telling them to get out. Walter called the nearby Wallace Fire

Department for help, and firefighters there loaded their private boats on trailers and headed down route 41. A neighbor showed up with a jet ski. Between Gary Cantrell's truck, the motor- and rowboats and the jet ski, 600 people were evacuated by daylight, when the firehouse had to be abandoned.

"Water rose to the roof of some houses. The current was so strong we had trouble maneuvering the boats . . . Our floodplain here is thirteen feet. We had twenty-five feet of water," Walter said.

"I felt pretty bad. I saw things I never thought I'd see. I saw the biggest snakes I'd ever seen on porches. Water moccasins. Rattlesnakes. Possums. Deer. Submerged vehicles everywhere. You'd be riding along in a boat and you'd hit something, and it would be a mailbox. There were dead chickens. Drowned wildlife. All the farms around here have diesel fuel tanks and the gasoline had spilled out."

In a post-greenhouse world, with the number of climate disasters rising, victims will be left more to their own devices. In North East, the National Guard would arrive later to cordon off the area. Federal and state legislators would tour the wreckage, shake Walter's hand and promise aid. But on that first night, the community was alone.

By the time Hurricane Irene was coming, aid was trickling in, but in ways that did not always help. The federal government offered loans for rebuilding houses, a debt burden beyond the reach of poorer folks in North East.

"How am I going to repay a loan?" said 80-year-old Earl Cavanaugh, whose 12,000 turkeys had drowned, whose soybean fields were covered with mud and sludge, and who could not conceive of taking out loans at this point in life. "I'd rather my house burned down. At least I'd have insurance. I don't know whether to bulldoze it or burn it."

Gary Cantrell said, "Two years ago my brother borrowed $40,000 to rebuild his home. Now his home is ruined, he still owes $40,000 and the bank wants to lend him more."

Bobby Hanchey, owner of the grocery, or what remained of it, said, "I watched $65,000 of stock go down the drain. Then the federal man says to me, I have to have proof. *Give me paperwork.* I pulled out paperwork. It was wet, stuck together . . . There are three hog lagoons near here, and there were 30,000 dead turkeys. I said to him, I ain't touching it. You go through it. He said, call me when it dries out."

Today, Bobby Hanchey drove around with a portable mosquito fog-ger, helping his neighbors ward off pests, as volunteers cleaned his store. There was nothing for North East to do but keep carting, scrub down walls with Clorox, pick up food at a free distribution center and keep the wives and kids with distant family until the homes were safe to reoccupy.

"The Red Cross asked me, what do I need?" said 71-year-old Ber-nice Cavanaugh, whose house had been ruined.

"I need a rich husband to buy me another home."

———

Back on the coast, an odd kind of detachment took hold of some peo-ple, while others experienced cumulative panic after so many storms.

"Normally I'd be at the supermarket, stocking up on supplies," one English professor at the university told a reporter in town. "But I'm in denial. I don't want it to be here. I'm not doing anything to prepare."

At New Hanover County Emergency Services headquarters at the Sheriff's Department in downtown Wilmington, phones rang off the hook with calls from residents asking if they should evacuate.

"This has got to *stop. It has just got to stop*," a receptionist called to a deputy, as they answered calls.

"We should be getting showers any time now," the deputy replied.

"*Last time* I-40 was underwater," one cop told another, as they came in from patrol.

"*Last time* we had a hole in that road, and a tractor trailer went through," the other cop said.

So many conversations were about "last time."

Out over the ocean, they all knew, Irene was a category 1 but that could change. She could hit warm ocean and strengthen. Her winds could grow nastier, her rain harder. She could turn the 500-year flood into a thousand-year flood. She could shut off electricity, rip roofs off churches, electrocute residents when they cleaned up. She could do enough damage to cost the university state funding, force it to hire fewer professors and delay building plans, all of which was ongoing after Hurricane Floyd.

After Irene, would there come a seventh hurricane? An eighth?

⊘ In Wrightsville Beach, by late afternoon, a humid breeze was picking up, and two red flags—hurricane warnings—snapped above town hall when Bob O'Quinn left the meeting of aldermen.

"In all probability we'll order the evacuation at eleven tonight, and start tomorrow at daylight."

He went home to get ready for the Centennial dance.

"The criteria will be, everyone off by thirty-five-mile-per-hour, gale-force winds. It's hard to transport equipment across the elevated bridge after that. We close down essential services when wind hits fifty miles per hour. But mandatory evacuation doesn't mean everyone has to leave. It means if you don't leave, you're confined to your property without essential services."

By 6:30, with O'Quinn getting into his tuxedo, out on the beaches it was still light, and a few joggers ran by the tideline, glancing at vastly contrasting views to the east and west. Inland, over the tops of the desirable beach homes, they saw the glorious pastel glow of a Carolina October evening, fiery, luminous, enchanting. But southeast, over a sea turning bottle green, the sky was a charcoal mass, with slanting lines of incoming rain spreading to engulf the horizon. It was like a curtain was being drawn across the sea—revealing the awesome and unparalleled beauty of a natural disaster seen from afar.

The first scattered drops tasted salty.

By nine, the gala was in full swing. Like revelers on the *Titanic,* celebrants danced to a live band, drank cocktails and dined on meat or fish amid decorations including potted palms and Christmas lights, thanks to Catherine O'Quinn, who looked smashing in a blue satin antebellum dress.

Bob O'Quinn wasn't drinking.

"We'll vote on evacuation at ten."

He wasn't having the best of times.

"We're waiting for the mayor and town manager to get back from a meeting at County Emergency Management."

Upstairs in his hotel room, with his brother and friends, he watched Irene on TV. She'd surprised everyone.

She'd sped up.

There would not be enough time to evacuate in the morning,

O'Quinn knew now. Evacuation, if it came, would have to be in the dark tonight.

At ten, O'Quinn went down to the ballroom and collected his fellow alderpeople. As jokesters among the revelers sang "Good Night Irene," drinks in hand, the alderpeople filed across the lobby to the reception desk, where a clerk led them to a conference room containing a large TV, broadcasting the hurricane, like every other TV set in town.

A commercial came on for electrical generators. It showed a teenage couple kissing during a power outage.

"Some people *like* the dark," the voice on the commercial said.

"Will someone turn that *off*?" an alderwoman snapped.

The police chief was at the table, and the fire chief. The mayor and town manager had just returned from the County Emergency Coordination meeting.

The mayor said, "We can expect gale-force winds at six A.M. Storm surge has been anticipated at four feet. Hurricane winds by four P.M. tomorrow. Landfall expected at six P.M. Winds expected at eighty to eighty-five miles an hour."

"By noon, we'll probably have lost power," an alderman said.

"If we don't restrict access, we'll have people trying to take their boats off with big rigs," said another

"I suggest we shut it down at gale-force winds."

"We're in complete denial," the alderwoman said.

The conversation picked up.

"The county is urging churches not to hold services tomorrow."

Evacuate or not? "Narrow window."

"I don't think we ought to do anything in the dark. It confuses people."

They voted and filed out of the conference room. In the ballroom, the music fell silent as Mayor Avery Roberts stepped up to the mike. He was a dignified-looking, white-haired man with a soft southern accent and a polite manner.

"We're not going to evacuate," he announced.

A rebel cheer went up, the kind that had been heard in Wilmington during the Civil War, when news of a Robert E. Lee victory arrived.

"We think this is the right thing to do," Roberts said.

O'Quinn hoped so. He really hoped so. The band swung into a last 1940s tune, a Tommy Dorsey number. The night was wearing on, as was the century that had worsened the natural greenhouse effect. By eleven celebrants were leaving, heading home, planning for the coming storm.

Bob O'Quinn ended the day as he had begun it, looking at the heavens, this time from his hotel room.

"We're taking a calculated risk," he said.

He was talking about evacuation, but on that night, with the greenhouse century a mere two months away, and fumes from carbon dioxide and methane and other greenhouse gasses rising—floating up into the atmosphere from millions of exhaust pipes and chimneys . . . on that night, with trillions of lightbulbs burning and air conditioners running around the globe, with the earth's population approaching 6 billion people, all of whom wanted a better life, heat when it was cold outside, a television to watch, a clean shirt, a computer to learn from, a real toilet, a hot-water tap, new glasses and any object requiring energy to manufacture—not bad people, *average* people who saw things and wanted them *and needed more energy*, and just didn't want to believe that the energy they wanted could change the atmosphere of earth, didn't even want to aggressively prepare for the possibility—on that night Bob O'Quinn also summed up the hopes of a planet.

He wasn't elegant at the moment, but he was accurate, with the hurricane approaching.

"If we guessed right, we'll be heroes. If not, we're screwed."

The Maldives

B Y NATIONAL PALACE STANDARDS, the presidential palace of Maumoon Gayoom was modest, taking up less space than a mansion in Beverly Hills. Enclosed by protective walls of whitewashed concrete, the house and courtyard combined Indian, Indonesian and Arabic elements, as befitting the home of the ruler of a country sitting astride trade routes.

The guards were tough, discreet and polite. Even when Gayoom went out, his security was lighter than the mayor of Atlanta's. Then again, he seemed more popular.

It was shortly after the turn of the twenty-first century, a fine, warm Indian Ocean night, and Gayoom had invited a reporter to the palace to talk about the greenhouse effect. The moon hung over Malé's low roofs and palms, the atoll, the nearby National Security headquarters, gold-leaf-roofed mosque, national museum and Sultan's Park, and Malé's vegetable and fish markets. The teahouses were open. The sea drive was alive with strollers and shoppers. Half a mile off, along the shore that had been topped by strange waves in 1987, a defensive sea wall was going up with Japanese aid. The leveelike leviathan of intertwined "tetrapods"—spiked granite blocks—fit into each other like gigantic jacks or petrified three-toed dinosaur feet.

All of the island would disappear if oceans rose.

More seawalls were being built to shield other islands, and the government had started moving people off sparsely populated ones, "consolidating our population on fewer islands," the country's environmental minister, Ismail Shafeeu, had told the reporter earlier that day. "If we require protection from storms or sea level change we'll be better off that way."

In the three-story palace, Gayoom's ground-floor meeting room was beautifully decorated by a deep Iranian carpet, comfortable couches and double doors of polished teak, their arches carved with Indonesian designs. The floor was white marble inlaid with onyx and black marble. The windows, framed in teak, were tinted blue.

At nine, the president walked in and led the reporter to an open-air courtyard, where, under the stars, servants had laid out soft drinks, small cakes and an array of samosas and spicy snacks of freshly caught tuna.

The talk was less pleasant though. "We are getting more frequent storms, stronger winds," the president said.

Gayoom talked about his innocent boyhood when, with his father, he sailed on wooden dhonis to visit far islands.

"In those days I did not realize that storms could kill people. Now I do."

He said that new kinds of erosion were eating away at 30 percent of the inhabited islands in the archipelago.

"People here tell me that during the last twenty years they have been losing land," he said, adding that historically, until now, ocean currents had shifted sand back and forth to islands in cycles, replacing land sucked from one side during the northwest monsoon season with sand deposited on the other side when currents switch to the southwest.

"It is hard for me to explain to people far away how our islands work," the president said, his frown conveying that distance from the rest of the world.

Gayoom had personally taken the reporter to one island earlier in the week, shown him places where trees and land had fallen into the sea, and then the two men had sat together as island fishermen set up chairs beneath the palms, brought coconuts as refreshment and told the president that their island was shrinking.

"We need a longer seawall," they said.

Now, in the palace, the president said politely, "Because of international arguing, if no immediate action is taken there might come a day when we are badly hit."

In the U.S., that night, events were, as usual, taking place that would affect the Maldives. A presidential election approached and candidates Al Gore and George W. Bush disagreed on climate change. To listen to them debate was to enter a time warp and hear positions taken 12 years earlier by Gore and another Bush, George W.'s father.

Gore would say, in the debate, "The world's temperatures are going up. Storms are becoming more violent and unpredictable . . . I'm strongly committed to . . . cleaning up new kinds of challenges like global warming . . ."

Bush would say, "I don't think we know the solution to global warming yet and I don't think we've got all the facts before we make decisions. I'll tell you one thing. I'm not going to let the United States carry the burden for cleaning up the world's air . . . Some of the scientists . . . haven't they been changing their opinion a little on global warming?"

Ironically, he would be referring to, of all people, Jim Hansen, still making headlines after over a decade of climate controversy. Expert reverses himself on global warming, news reports had announced, then asserted that Hansen now believed the warming might *not* turn out as severe as he originally predicted.

And Hansen, distressed at the misinterpretation of his newest scientific paper, would throw himself into interviews to correct it. He had *not* changed his mind, he'd insist. He *still* believed warming was dangerous. He had merely tried to suggest that damage might be mitigated faster if countries cut emissions of other greenhouse gasses before carbon dioxide. The process would be cheaper, and the effects felt sooner, although CO_2 would eventually also have to be reduced, he said.

"I was trying to say it would be possible to slow global warming and improve public health by reducing emissions of soot from diesel engines, and from coal," Hansen said, upset.

But the clarifications would never be as widely disseminated as the first flawed report, and Hansen, of all people, would become ammunition for the GCC in 2000. "See," fossil fuel lobbyists would tell reporters at every opportunity, grins on their faces. "The guy who started this whole ridiculous mess has *changed his mind!*"

In the Maldives, though, with the tropical moon up, and the cool sea breeze bathing the president's courtyard, it was difficult to keep talking about how everything in view could disappear beneath the sea. So at length the talk shifted to Gayoom's plans for the country if *no* global warming destroyed it, and turned his palace into a reef, his home into driftwood, his people into floating refugees.

Gayoom welcomed the opportunity to discuss happier things and smiled while talking about his "2020 Project."

"It is what we think the country will be like in 2020, *providing there will be no ocean rise* . . . We want to provide electricity at an affordable price and safe water to all the islands," he said, sounding like any developing-country president, who *needed more energy*. "At the moment, we have electricity, but not enough. We want to have drinking water available for all the islands. We want each and every person on every island to have access to health care . . . We want to upgrade our educational system."

Listening to him, the reporter was struck once again by the schizo-phrenic dance between hope and despair that he had encountered among Maldivean leaders.

On one hand, everywhere in the capital was vigorous progress. Health care was better. Schools were better. Dredging projects abounded. There were plans to enlarge the airport, and a national tree-planting program was under way.

Even the reefs, damaged by high sea temperatures in 1998, were thriving again. Tourism was booming. New resorts were planned.

Day to day, life remained slow and pleasant.

But at the same time, with storms and winds worsening, on the rare days they came, Australian experts in the Maldives were working to identify coral reefs which were less susceptible to higher ocean temper-atures and to ultraviolet radiation. These pre-adapted reefs would receive federal protection, as they were more likely to survive the future.

Reef triage had begun in the Maldives.

In another, U.S.-sponsored project, on an island an hour from Malé by fast boat, an automated tower filled with weather-measuring equip-ment—part of a system of stations scattered through the Indian Ocean—recorded a steady stream of data sent back to the U.S. to help researchers like Tom Karl's monitor the atmosphere. And that information sup-ported conclusions that the atmosphere was deteriorating.

"You can draw parallels between us and the threat that America faced during the nuclear missile crisis," Minister Shafeeu had told the reporter, when asked about the regular shift between hope and despair. "Back then, if you tried to talk to Americans about atomic bombs, they did not want to discuss it, but every newspaper, magazine, and every James Bond movie was about it. And at that time, like the carbon dioxide now, the whole thing was a *system*. You had factories producing missiles. The economics were tied into it. *Everything* seemed tied into it . . . Some general in Russia who has no link with you pushes a button and becomes responsible for doing something that destroys your life."

Back and forth. Apprehension to hope. "We believe in Providence," President Gayoom said at one point. "We don't really think that the country will drown. We want to believe there is still time worldwide to save the Maldives and other low-lying countries."

The reporter pointed out that this statement didn't match up with direr predictions the president had made at international meetings, and Gayoom nodded that this was true, but when at home it was important not to alarm people.

Or was something else happening here? the reporter wondered. By that night, he had heard critics of the Kyoto treaty often charge that third world politicians merely used it as an excuse to get aid from developed countries. They didn't really believe that the climate was changing. They simply used it to beg for cash. Was it possible there was some truth in the charges? Was *that* why the president seemed to be saying two opposite things?

But then the reporter remembered something the Minister of Environment had told him.

"For those of us trying to work on this thing, that sense of impending doom all the way through our lives on these islands—it would be very difficult if we let it in. If you're not involved day to day in reading the documents that tell you that you are going to be gone . . . if you don't see it all the time, maybe you can put your mind away from it and say, there are people working to change it. Maybe I don't have to worry about it," Ismail Shafeeu had said.

"But when you're in a situation where daily you come in contact with statements and newspaper reports saying that in fifty years you might not be there, you have to *consciously* not think about it or you totally give up. We have to retain some element of optimism. Some element of hope."

President Gayoom was trying to retain hope. He needed to keep from becoming like "some people," as Shafeeu had said, "like a Jamaican I know involved with environmental NGOs. He, I think, is going crazy. He has come here, trying to come up with ways of encouraging coral growth, protection for the islands, and over time he has been getting more emotional. We met him in New York recently, at the UN, and he had his kid with him. The poor child. The guy was going on and on, with charts, talking about warming trends in the Indian Ocean. Telling us, *Can't you see what's happening to your country? Can't you see what's happening to your reefs?*

"I couldn't talk to him. A lot of what he said was true, but you get to where you think, what's the point of talking? Nobody's going to do anything in time to save your country. You get into that state of mind where all you think about is survival rather than trying to change things. But at least now we still keep in our minds the *possibility* that maybe we can make a difference. You have to have some optimism or otherwise you totally let go."

And now, at the palace, it was growing late, in the evening, and in the history of Gayoom's nation.

Within months, George W. Bush would win the presidential election in the U.S. He'd repudiate the Kyoto treaty. He'd push for the opening up of public lands for *more* oil exploration. He'd launch a full-scale effort to respond to an energy crisis inside the U.S. by promoting greater reliance on fossil fuels, without undertaking any substantial effort to wean the country off traditional energy policy, or gluttony, environmentalists would charge.

"When you talk to me," the reporter said to Gayoom, "you're not talking to people in the Maldives, whose spirits you want to keep up. They're not here with us."

The president sat back, exhausted from the hour.

The reporter asked if there were active plans to evacuate the nation.

"No," Gayoom said.

He asked if it was time to start drawing them up.

Gayoom hesitated. He sighed.

He said, in a voice so soft it was almost a whisper, "Yes."

Karl, Tom, 45, 46–53, 59, 61, 64, 68, 89,
  92–93, 104, 147, 170, 209, 262, 308
  position on global warming, 176–77,
    204–5, 210, 277–81
Kimball, Bruce, 123
Kohl, Helmut, 128
Koppel, Ted, 172–76, 212, 214
Kozel, Petr, 244
Kunisakera, Varuni, 272–76
Kyoto conference/summit, 193, 224, 239, 243,
  246–51, 287
Kyoto Protocol, 250–51, 284, 288, 290
  and environmental security, 245–46
  Kyoto treaty, 287, 309
  repudiated by Bush, 310

La Niña, 277
Lawrence Livermore National Laboratory,
  171, 206, 207, 215
Leggett, Jeremy, 66, 120, 124, 285
Levee system/levees, 155, 157, 160, 161, 166,
  167, 168
  rebuilding, 164–65, 168–69
Lichtblau, John H., 59
Lieberman, Joe, 265
Lietzmann, Kurt, 243–44
Limbaugh, Rush, 124, 174
Lindzen, Richard, 61, 64
Lobbying/lobbyists, 4, 170
  corporate/industry, 61, 62–63, 119, 122,
    211, 212
  against treaty for Kyoto, 243, 246
  in treaty negotiations, 124–25, 243
Lockheed, 264
London, storm in, 83
Longinotti, Kevin, 4–7, 12, 14–15, 17–18,
  19–21, 279
Los Angeles, 93, 289
Low-pressure areas, 67, 69

McDonald, Emily, 71, 78, 79, 80–83, 85, 86,
  90, 91, 228, 279
McGinty, Kathleen, 148–50, 151–53, 154, 163,
  165, 170, 240, 246–47, 248–49, 251
McInnis, Michelle, 198–201
Mack, Connie, III, 265
McLure, James A., 31
Madrid meeting, 204, 205–6, 208–9, 213,
  215
Majeed, Abdullahi, 97, 100, 220–21, 248
Malaria, 27, 273–74, 288, 292
Maldives, 94–102, 119, 133, 190, 203–4, 209,
  220–21, 305–10
  death of coral reefs in, 257, 258
Malé, 94, 98–99, 100, 101, 305

Mammoth, Wyoming, 32, 39, 40
Mann, Michael, 252–56, 278
Mansley, Mark, 186–87
Marks, Frank, 138, 139–40, 142–43, 146
Mars, 24
Martin, Derrel, 15–16, 19
Martin, James, 75–77, 83–85
Martinez, Bob, 60
Mass migration, 4, 245, 246
Mathur, Ajay, 190, 281–84
Matthew, Richard, 234–38, 243, 244, 245
Mauna Loa Observatory, 26
Max Planck Institute for Meteorology, 105,
  205, 206
Mayeux, Herman, 123
Maytag, 264
Meltponds, 261
Memphis, Tennessee, 25, 30
Mercury barometer, 66
Mesocyclone, 11, 13
Meteorologists, 10, 18–19, 25, 93, 142
  in Britain, 69
Methane, 169, 247, 304
Methane hydrates, 125
Miami, Florida, 60
  Hurricane Andrew, 141–42
Michaels, Patrick, 61–62, 64–65, 174, 176,
  203, 241
Mideast, 237
Midwest, 93, 155
  floods, 168
Minnesota, 25, 156
Mississippi River, 25, 155, 156, 157, 159, 160,
  167
Missouri, 209
  floods, 156–69
Missouri River, 155, 157, 158–59, 160,
  161
MIT, 26, 171
Mobil, 27, 125, 224
Models
  see Climate models; Computer models
Molecules, 66–67
Montgomery County (Tennessee), 10
Moon, Sun Myung, 173, 174, 175, 176
Moore, Thomas Gale, 203–4
Mosquitoes, 272–73, 274–75
Mt. Pinatubo, 102–4, 105, 107, 177
Munich Re, 257
Murkowski, Frank, 32, 192
Murphy, Mike, 13

Nagasaki, 8
Nana (storm), 85–86
NASA, 23, 58, 219, 257